Energy Efficiency Guide
for Existing
Commercial Buildings:
Technical Implementation

This is an ASHRAE Design Guide. Design Guides are developed under ASHRAE's Special Publication procedures and are not consensus documents. This document is an application manual that provides voluntary recommendations for consideration in achieving greater levels of energy savings relative to minimum standards (code).

This publication was prepared under the auspices of ASHRAE Special Project 125 and was supported with funding from DOE through PNNL contract #126760.

ENERGY EFFICIENCY GUIDE FOR EXISTING COMMERCIAL BUILDINGS—TECHNICAL GUIDE PROJECT COMMITTEE

George Jackins, *Chair*

Tom Watson
ASHRAE Representative

Ron Burton
BOMA Representative

Kevin Kampschroer
GSA Representative

Kinga Porst
GSA Alternate

Rita Harrold
IESNA Representative

Brendan Owens
USGBC Representative

Adrienne Thomle
C 7.6 Representative

Harry Misuriello
TC 7.6 Representative

Gordon Holness
Member-At-Large

Jim Bochat
Member-At-Large

Bruce Hunn
Member-At-Large

Bert Etheredge
ASHRAE Staff Liason

Lilas Pratt
ASHRAE Staff Support

AEDG STEERING COMMITTEE

Don Colliver, *Chair*

Bill Worthen
AIA

Costas Balaras
ASHRAE

Rita Harrold
IESNA

Brendan Owens
USGBC

George Jackins
EEG-EB

Merle McBride
AEDG

Paul Torcellini
AEDG

Lilas Pratt
Staff Liason

Jeremy Williams
U.S. DOE

Adrienne Thomle
Consultant (ASHRAE TC 7.6)

Michael Schwedler
Consultant (ASHRAE Std. 90.1)

Kent Peterson
Consultant (ASHRAE Std. 189.1P)

Ron Jarnagin
AEDG

Shanti Pless
AEDG

Bing Liu
AEDG

Terry Townsend

Energy Efficiency Guide for Existing Commercial Buildings: Technical Implementation

Lead Author
Dennis R. Landsberg, L&S Energy Services, Inc.

Contributing Authors
Steven Carlson, CDH Energy Corp.
Fredric S. Goldner, Energy Management & Research Associates
J. Michael MacDonald, Energy Performance Measurement Institute
Ronald B. Slosberg, L & S Energy Services, Inc.

Developers
American Society of Heating, Refrigerating
and Air-Conditioning Engineers, Inc.
The American Institute of Architects
Illuminating Engineering Society of North America
U.S. Green Building Council
U.S. Department of Energy

Contributors
Building Owners and Managers Association International
U.S. General Services Administration

ISBN 978-1-936504-17-6

© 2011 American Society of Heating, Refrigerating
and Air-Conditioning Engineers, Inc.
1791 Tullie Circle, NE
Atlanta, GA 30329
www.ashrae.org

Cover design by Emily Luce, Designer.
Cover photographs courtesy of H.E. Burroughs

Library of Congress Cataloging-in-Publication Data

Landsberg, Dennis R., 1948-

Energy efficiency guide for existing commercial buildings : technical implementation / Lead Author Dennis R. Landsberg ; contributing authors Steven Carlson... [et al.].

p. cm.

Includes bibliographical references and index.

Summary: "This guide explains why building owners and managers should be concerned about energy efficiency. The second in a series, the goal of this resource is to help the audience evaluate current building operations and also provides technical guidance on how to achieve energy efficiency in these buildings"--Provided by publisher.

ISBN 978-1-936504-17-6 (softcover : alk. paper) 1. Buildings--Energy conservation. 2. Commercial buildings--Energy consumption. 3. Commercial buildings--Cost of operation. I. Title.

TJ163.5.B84L354 2011

658.2'6--dc23

2011043253

ASHRAE STAFF

SPECIAL PUBLICATIONS

Mark S. Owen
*Editor/Group Manager
of Handbook and Special Publications*

Cindy Sheffield Michaels
Managing Editor

James Madison Walker
Associate Editor

Elisabeth Parrish
Assistant Editor

Meaghan O'Neil
Editorial Assistant

Michshell Phillips
Editorial Coordinator

PUBLISHING SERVICES

David Soltis
*Group Manager of Publishing Services
and Electronic Communications*

Jayne Jackson
Publication Traffic Administrator

PUBLISHER

W. Stephen Comstock

**Any updates/errata to this publication will be posted on the ASHRAE
Web site at www.ashrae.org/publicationupdates.**

Contents

Acknowledgments

The *Energy Efficiency Guide for Existing Commercial Buildings: Technical Implementation* addresses the existing commercial building stock in the United States and provides technical guidance on how to achieve energy efficiency in these buildings.

The main contributors to the Guide were the primary author, Dennis R. Landsberg, and the contributing authors, Steven Carlson, Frederic S. Goldner, J. Michael MacDonald, and Ronald B. Slosberg. These authors put in many long hours and worked collaboratively to produce this Guide.

The Advanced Energy Design Guide (AEDG) steering committee provided direction and guidance to complete this manuscript and produced an invaluable scoping document to begin the creative process. Members on the project monitoring committee came from the AEDG steering committee partner organizations as well as the Building Owners and Managers Association and the U.S. General Services Administration. These members served not only on the project monitoring committee but also as liaisons to their respective organizations.

The Chair would like to personally thank the authors and all the members of the project monitoring committee for their diligence, creativity, and persistence. The Chair would also like to thank Michael Jouaneh for his contributions to the lighting measures.

In addition to the authors and the members on the committee, there were a number of other individuals who contributed to the success of this Guide. The specific individuals and their contributions were: Lilas Pratt and Bert Etheredge of ASHRAE for their assistance, organizational skills, and dedication to the project; and Meaghan O'Neil of ASHRAE

Special Publications for editing and laying out this book. This Guide could not have been developed without all of their contributions.

I am very proud of the Guide that has been developed and amazed at the accomplishment in such a short time period. The authors and each of the project monitoring committee members should be proud as well of their individual contributions to this most worthwhile document.

George Jackins
SP-118 Chair
October 2011

Abbreviations & Acronyms

A/C	=	air conditioning
AEDG	=	*Advanced Energy Design Guide*
AEE	=	Association of Energy Engineers
AIA	=	American Institute of Architects
ANSI	=	American National Standards Institute
ASHRAE	=	American Society of Heating, Refrigerating and Air-Conditioning Engineers
BEAP	=	Building Energy Assessment Professional
bhp	=	brake horsepower
BOMA	=	Building Owners and Managers Association International
Btu	=	British thermal unit
CBECS	=	Commercial Buildings Energy Consumption Survey
CCE	=	cost of conserved energy
CE	=	combustion efficiency
CEA	=	Certified Energy Auditor
CEM	=	Certified Energy Manager
CF	=	cubic feet
CHWST	=	chilled-water supply temperature
CO_2	=	carbon dioxide
CRI	=	color rendering index
CWST	=	condenser-water supply temperature
DCV	=	demand control ventilation
DD	=	degree day
DDC	=	direct digital control
DHW	=	domestic hot water
DOE	=	U.S. Department of Energy
dP	=	differential pressure
DSIRE	=	Database of State Incentives for Renewable Energy
DSP	=	duct static pressure
ECI	=	energy cost index
EEG-EB	=	*Energy Efficiency Guide for Existing Commercial Buildings*

EEM	=	energy efficiency measure
EEPS	=	energy efficiency portfolio standard
EER	=	energy efficiency ratio
EIA	=	Energy Information Administration
EMCS	=	energy management and control system
EMS	=	energy management systems
EPA	=	U.S. Environmental Protection Agency
ESCO	=	energy service company
ESPC	=	energy saving performance contracting
EUI	=	energy utilization index
FEMP	=	Federal Energy Management Program (developed by the National Institute of Standards and Technology)
gpm	=	gallons per minute
GSA	=	U.S. General Services Administration
HID	=	high-intesity discharge
HVAC	=	heating, ventilating, and air-conditioning
IAQ	=	indoor air quality
IES	=	Illuminating Engineering Society of North America
IPE	=	industrial process efficiency
IRR	=	internal rate of return
IPMVP	=	International Performance Measurement and Verification Protocol
kW	=	kilowatt
kWh	=	kilowatt-hour
LCC	=	life-cycle costing
LED	=	light-emitting diode
LEED—EB	=	Leadership in Energy and Environmental Design —Existing Buildings
LF	=	load factor
low-e	=	low-emissivity
M&V	=	measurement and verification
MARR	=	minimum attractive rate of return
MERV	=	minimum efficiency reporting value
NAESCO	=	National Association of Energy Service Companies
NEMA	=	National Electrical Manufacturers Association
NPV	=	net present value
NYSERDA	=	New York State Energy Research and Development Authority
O_2	=	oxygen
OA	=	outdoor air
occ	=	occupied
OIT	=	DOE Office of Industrial Technology
OM&M	=	operations, monitoring, and maintenance
P.E.	=	professional engineer
PM	=	Portfolio Manager
PMP	=	performance measurement protocols
ppm	=	parts per million

RF	=	radio frequency
RFP	=	request for proposal
ROI	=	return on investment
RSC	=	Roof Savings Web Site Calculator
SAT	=	supply air temperature
SBC	=	systems benefits charge
SEER	=	seasonal energy efficiency ratio
SHW	=	service hot water
SHGC	=	solar heat gain coefficient
SIR	=	savings-to-investment ratio
Std Dev	=	standard deviation
USGBC	=	U.S. Green Building Council
VAV	=	variable air volume
VFD	=	variable-frequency drive
VSD	=	variable-speed drive
VT	=	visible light transmission
W	=	watts

Preface

This guide is the second in a series of planned Energy Efficient Guides for Existing Commercial Buildings (EEG-EB) developed by the American Society of Heating, Refrigerating and Air-Conditioning Engineers (ASHRAE) in collaboration with the American Institute of Architects (AIA), the Illuminating Engineering Society of North America (IES), and the U.S. Green Building Council (USGBC), and supported by the U.S. Department of Energy (DOE). While the first guide in the series provided the business case for improving energy efficiency, this guide provides technical guidance on how to increase a building's energy efficiency.

This guide is intended for building engineers and managers and assumes a working technical knowledge of building systems both generally and for the reader's specific building. The guide demonstrates ways to measure a building's energy efficiency, track that efficiency, develop an energy efficiency plan, and provides guidance on implementing the developed plan. The goal is to provide clear and easily understood technical guidance for energy upgrades, retrofits, and renovations, by which building engineers and managers can achieve at least a 30% improvement in energy performance relative to a range of benchmark energy utilization indexes (EUIs). The document provides practical means and methods for planning, executing, and monitoring an effective program, based on widely available technical strategies and technologies.

ASHRAE has a number of standards, guidelines, and design guides that provide guidance for constructing energy efficient buildings. This new guide dovetails well with these documents; yet, realistically, these other efforts primarily target new construction or, to a limited extent, portions of existing buildings undergoing major renovation.

The highly successful Advanced Energy Design Guide (AEDG) series provides prescriptive guidance for new building designs to achieve energy savings 30% to 50% beyond ANSI/ASHRAE/IES Standard 90.1. The building types covered by these guides include small to medium office buildings, retail buildings, K–12 school buildings, warehouses, highway lodging, hospitals, and healthcare buildings.

Unlike the new building guides, which were written by teams of volunteers from the collaborating organizations, the existing building guides are written by a contractor, under the guidance of an EEG-EB Project Monitoring Committee. In addition to the collaborating partner organizations, the Building Owners and Managers Association (BOMA) International and the U.S. General Services Administration (GSA) were also involved in the development of the guide and represented on the Project Monitoring Committee.

The existing building guides also address a wider range of issues than the new building guides. While the new building guides addressed only design issues, the existing building guides address, at some level of detail, analysis of current operations, renovation, retrofit, system or equipment replacement, operation and maintenance, as well as providing technical design input. These issues will be strongly dependent on the building type and age. And, virtually all existing buildings will first need to be brought up to minimum requirements before energy savings are pushed beyond minimum performance. A hierarchical approach based on the cost of conserved energy provides the basis for prioritized measure selection is presented in the context of life-cycle costing that encourages long-term investment horizons.

Clearly, the greatest opportunity for overall reduction in U.S. primary energy use lies within the existing building stock. That stock also represents a significant potential for real estate owners and developers to not only demonstrate sustainability initiatives, but also to realize a great return on the investment. This series of EEGs, in addition to other ASHRAE initiatives aimed specifically at promoting energy conservation and efficiency in existing buildings, will help building owners achieve the goal of building sustainability while increasing their return as energy prices rise.

Ronald E. Jarnagin

ASHRAE President

October 2011

Introduction 1

This guide is the second in the existing building series and is presented primarily for use by building managers and operators, and as a reference for owners. Its purpose is to explain technical implementation of energy efficiency in existing commercial/institutional buildings or at campuses or complexes. The first guide, *The Business Case for Building Owners and Managers*, articulated the rationale for building energy efficiency (Landsberg et al. 2009). In this guide, commitment by the reader to pursuing improved energy efficiency is presumed, and the focus is on implementation. The reader will learn how to measure the relative energy efficiency of a building, track that efficiency over time, and develop and implement an energy efficiency plan for the building.

Energy efficiency improvement in buildings is one of the greatest means to increase resource efficiency, improve environmental stewardship, and save operating funds. More importantly, *energy efficiency improvement should happen because it makes good business sense.* Good planning and ongoing commitment are essential to maximizing investments in energy efficiency.

BASIC ENERGY EFFICIENCY PROCESS

The overall basic energy efficiency process is shown in Figure 1-1 in the energy efficiency process flow diagram. The process of achieving energy efficiency improvements in commercial buildings can be handled from two main starting points. The two colored boxes at the top of the figure show the performance calculation and goals-setting steps recommended. The gray boxes show the process that has been

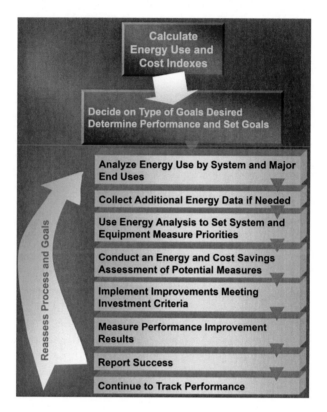

Figure 1-1. Energy efficiency process flow diagram.

commonly known as an *energy audit* or *energy assessment*, with additional performance tracking steps at the end. It must be stressed that this is an iterative process. Continual monitoring and reassessment (operations, monitoring, and maintenance—OM&M) of the building energy use, ongoing commissioning, and periodic reauditing are essential to keeping the building both current technologically and in tune.

The initial performance and goals-setting steps should be completed in order to better understand the savings potential and savings already achieved, but a basic energy assessment can also be conducted to identify and implement energy efficiency improvements, increase efficiency, and save energy.

MEASURING HISTORICAL ENERGY PERFORMANCE AND SETTING GOALS

The first guide in this series, *The Business Case for Building Owners and Managers*, presented the case for measuring energy efficiency and setting goals (Landsberg et al. 2009). In some locations, performance measurement is now mandated with increasing coverage of the building stock over time. The best understood means of measuring performance is to calculate the whole-building energy utilization index (EUI) and energy cost index (ECI), as laid out in ANSI/ASHRAE Standard 105-2007, *Standard Methods of Measuring, Expressing and Comparing Building Energy Performance* (ASHRAE 2007b). This calculation method will be described in simplified form in Chapter 3 so that basic performance can be calculated.

For more detailed measurements the reader is referred to ASHRAE's performance measurement protocols (PMP) for commercial buildings (ASHRAE 2010e). The PMP document presents protocols for the energy performance of commercial buildings at three levels of detail: basic, intermediate, and advanced, whereas the EEG—EB (*Energy Efficiency Guide for Existing Buildings*) document addresses measurement only at the basic (whole-building monthly and annual) level; i.e., EUI and ECI (Landsberg et al. 2009). The PMP document provides guidance for measuring performance at the system or end-use level, and using weekly, daily, or hourly data, which are not covered here.

For some jurisdictions, it may be necessary to calculate carbon emissions or other energy-derivative values, such as tons of carbon emitted or saved, but these types of values will not be considered directly in this book. The energy values are needed in order to calculate the energy-derivative values. Achievement of some types of certification of energy efficiency, such as LEED—EB (Leadership in Energy and Environmental Design—Existing Buildings) (USGBC 2008), or environmental performance, such as EUI or carbon reduction, may also be of interest for some building owners. For example, a jurisdiction may set energy targets for specific building types that must be met before the building can be sold. The energy-use data are needed to apply for such certifications as well.

USING CONSULTANTS OR ESCOS

In some instances, a consultant may be hired to handle many or all the steps of the process, but the basic need is to identify key activities, designate who will be responsible for carrying out each activity, and define how progress will be tracked and measured. Owners of small buildings may have to rely entirely on outside parties for all aspects of the process, except making the commitment. When outside parties are involved, building trust is required for each relationship (a Business Relationship Trust Equation is presented at: http://trustedadvisor.com [Green 2007]). For larger buildings or facilities, an energy service company (ESCO) can turnkey the entire process, with direction from the owner, including providing project financing.

A commissioning agent, either for new equipment installation or to recommission existing systems, is another consultant service that should be strongly considered. Without proper commissioning, even the best and most efficient systems can run in an inefficient manner—resulting in the anticipated energy savings not being achieved.

More information on using consultants or in-house staff is presented in Chapter 2 under "General Considerations." Chapter 2 also describes initial considerations for using an ESCO.

IDENTIFYING AND IMPLEMENTING ENERGY EFFICIENCY MEASURES

Chapter 3 presents a methodology for calculating building energy use and cost. Setting energy efficiency goals is also discussed.

Chapter 4 of this book provides information on how energy use is analyzed to identify potential energy efficiency measures. Chapter 5 provides methods for calculating savings for common energy efficiency measures and how to select measures for specific buildings. Chapter 6 presents methods for refining financial analysis and setting priorities on which measures to implement. The concept of life-cycle costing is discussed. Chapter 7 brings the whole process together and discusses measurements needed to verify savings, why reporting success can be important, and the importance of continuing to track energy performance.

OWNERS OF SMALL AND MEDIUM SIZE BUILDINGS

While this guide is written for all commercial buildings, much of the material pertains to large buildings. As such, it contains material suited to large buildings, which is too complex for use by the owners of small buildings. Section 2 contains a subsection designed to assist owners and managers of small buildings in using the guide.

INDOOR ENVIRONMENTAL QUALITY

There may be unintended effects on the indoor environment from energy efficiency retrofit activities. While the scope of this guide does not include potential indoor environment impacts, readers should be aware that changes to the building can have effects related to noise, lighting, breathing air (indoor air), and temperature- and humidity-related comfort. Concerns about noise and lighting can usually be handled by asking simple questions, such as:

- Will this change cause any new noises that may bother us (or the occupants)?
- Will this new equipment be quieter than what we have?
- If there are noises after you make the change, can you come and fix them?
- Can you show us what this new lighting will look like (asking for a demonstration)?
- How quickly can the new lights turn ON after they are turned OFF?
- How often can the lights be turned ON and OFF?
- Who will fix the lighting controls if they turn ON when they are not supposed to?

Temperature- and humidity-related comfort can be complicated to discuss, but simple questions should still be asked of retrofit providers, such as:

- What will the temperature be like after this change is made?
- How well does this new equipment keep the humidity down in the summer?
- How does this equipment keep things from being too dry in the winter?

For breathing air (indoor air), the topic is usually called indoor air quality (IAQ). ASHRAE has developed extensive guidance on indoor air quality, but mostly related to new construction: *The Indoor Air Quality Guide: Best Practices for Design, Construction and Commissioning* (ASHRAE 2009d).

For a free booklet on indoor air quality, a 200+ page Environmental Protection Agency book, *Building Air Quality: A Guide for Building Owners and Facility Managers* (EPA 1991), can be downloaded from www.epa.gov.

Chapter 2 in this book (EPA 1991), pages 5–12, provides some background information that would allow questions about indoor air quality to be asked before changes are made to the building.

MAINTAINING OCCUPANT COMFORT

Human comfort is another important consideration in embarking on an energy efficiency program. Energy efficiency should not be achieved by sacrificing occupant comfort and safety. ANSI/ASHRAE Standard 55-2010, *Thermal Environmental Conditions for Human Comfort* (ASHRAE 2010a) and ANSI/ASHRAE Standard 62.1-2010, *Ventilation for Acceptable Indoor Air Quality* (ASHRAE 2010b) are important references in maintaining occupant comfort.

General Considerations 2

Improving business/process efficiency, reducing operating costs, and increasing output and/or profitability is a goal of any commercial business, municipality, or other enterprise. However, some general considerations must often be taken into account before starting the process or during the process. In this chapter, an overview of some important general considerations is presented, such as limits caused by total annual energy costs, effects of owner/property types, information materials on resources, how to hire consultants, use of energy service companies, tenant effects, and utility rate issues.

Much of the material in this book is targeted at improving systems in existing buildings. However, for owners involved in new construction, the concepts are the same but the approach is different. With an existing building, energy efficiency improvements are constrained by what is already in place. Major equipment replacements often have long payback periods. Therefore, economics dictate that major improvements be tied to the life cycle of the building systems to leverage the replacement cost of equipment when making improvements. Since capital is necessary to replace the equipment anyway, the energy efficiency savings can be based upon the increased cost of more efficient equipment, rather than the total cost of the equipment. In new construction, all building systems are being purchased and installed. Upgrades can be justified based upon the incremental cost of the higher efficiency equipment as compared to code compliant equipment.

TOTAL ENERGY COST SHOULD DRIVE THE PROCESS

Any potential process improvement is first compared to its potential benefits before a decision is made regarding implementation. With regard to energy efficiency improvements, the starting point is existing total energy use and cost, where the total includes payments by both owners and tenants. Evaluating the total energy cost, or preferably the life-cycle cost of operating a building, and the costs relative to a "reference" or "typical" building EUI or ECI will give building managers, decision-makers, and owners an indication of the energy and energy-cost savings that may be achievable through the completion of an energy efficiency retrofit program. The greater the differential between the building's energy use and the reference or typical energy use, the greater the potential for energy efficiency and energy savings. Calculations and comparisons are discussed in some detail in Chapter 3, but total energy cost is the main value to examine before starting.

While energy cost drives the process, energy use is the yardstick by which progress is measured. Goals must be set and progress tracked based upon energy use, because energy costs change over time. As energy costs change, the relative priority of opportunities may change, and the energy efficiency plan needs to be updated.

Table 2-1, energy cost options, provides an example of how total annual energy cost sets limits on how much might be invested in energy efficiency measures for a building. These numbers are not meant to be "hard" or strict, but do provide a sense of scale for approaching energy projects in buildings. If total project investment also must include costs for outside consultants, the scale of potential investment becomes even more important. Use these values to help scale the potential scope of projects before extensive planning. Keep in mind that a potential project investment of $50,000 or more may be needed to provide enough resources for outside consultants.

Typical cost savings are in the 10% range, but can be as high as 20% or more. The potential project investment depends on how many years a building owner is willing to allow to achieve a positive cost benefit. If a quick financial analysis indicates that an owner would be able to see a cost benefit in eight years, the potential annual savings would be multiplied by eight to estimate approximate potential project investment. The rule-of-thumb savings is 10%–20% of energy costs for estimating project size.

Table 2-1. Energy Cost Options (in 2010 dollars)

Option Level	Annual Energy Cost, $/yr	Potential Project Investment, $
Low	1000–10,000	1500–10,000
Medium	10,000–30,000	4000–50,000
High	30,000–60,000	15,000–100,000
Very High	More than 60,000	30,000–Millions

Based on 10%–20% potential savings on energy cost, and an annual energy cost of $50,000/yr, the potential savings range is $5000–$10,000/yr. The values in Table 2-1 are roughly based on using five times possible annual savings for estimates for buildings with low-energy costs and up to ten times annual savings for buildings with high-energy costs.

IMPACT OF PROPERTY AND OWNERSHIP TYPES

Commercial business owners make financial decisions using different criteria than municipal and institutional managers. Profits and the cost of capital generally mean that commercial businesses will focus on internal rate of return, or short-term payback when making investment decisions. The return on investment of a dollar spent on energy efficiency will ideally be compared to a dollar spent on improving sales or profitability from nonenergy investments.

Institutional and municipal or other government decision makers generally have a more long-term focus. While short paybacks and high return-on-investment (ROI) are desirable, measures with longer payback periods are often adopted. While business owners often want paybacks of less than three to five years before making an investment decision, institutional and municipal/government decision-makers use a longer time horizon, often eight to ten years. The federal government uses savings-to-investment ratio (SIR) to make energy efficiency decisions. A SIR>1 means that an efficiency measure will pay for itself in a time period that is shorter than the life of the measure. Warranties and expected lives of equipment and/or systems factor in to these calculations.

ederal government has extensive regulations that govern energy assessment and economic analysis procedures. These requirements cannot be covered here, so managers of federal buildings will need to consult the appropriate regulations. State and local government organizations may also have regulations impacting these activities.

Since many businesses lease or rent their properties, the business often does not own the lighting, HVAC (heating, ventilating, and air-conditioning), and other building systems, yet they incur the energy costs for operating the equipment. The cost for energy efficiency upgrades would typically be incurred by the building owner, while the operating cost savings would benefit the tenant. This situation results in hesitancy from building owners to implement energy efficiency measures that would benefit the business owner that rents space in the property. It should be noted that as the competitiveness in rental real estate market increases, landlords will be increasingly pressured to invest in energy efficiency to reduce tenant operating costs. In many markets in the United States, the continued awareness and focus on sustainability and energy efficiency has begun to result in an increase in efficiency projects in lease and rental properties. The addition of tenant submetering rather than apportioning utility bills by space use will encourage tenants to limit energy consumption.

Many documents have been written on the impacts that different leasing arrangements have on energy efficiency activities, but there are too many to cover here. The most mentioned guide to writing a "green" lease is the BOMA (Building Owners and Managers Association) International *Green Lease Guide*, which walks users through the process of making leases "greener." The lease guide includes energy efficiency provisions. The *Green Lease Guide* was recently updated to: *BOMA International Commercial Lease: Guide to Sustainable and Energy Efficient Leasing for High-Performance Buildings* (BOMA 2011).

UTILITY ENERGY PROGRAMS AND RATE STRUCTURE EFFECTS

If the local or regional electric and gas (or other) utility company offers energy efficiency programs that have investment incentives, energy efficiency goal-setting should include the potential effects of the incentives when setting efficiency goals or planning efficiency improvement programs. Tables 2-2 and 2-3 are examples of utility and state energy efficiency programs.

**Table 2-2. National Grid Upstate New York Utility Rebate
Program—from DSIRE Database**

State	New York
Incentive Type	Utility Rebate Program
Eligible Efficiency Technologies	lighting, lighting controls/sensors, compressed air, energy management systems/building controls, custom/others pending approval, LED (light-emitting diode) exit signs, vending machine controls, commercial refrigeration equipment, commissioning, technical assistance, hotel occupancy sensors
Applicable Sectors	commercial, industrial, nonprofit, schools, local government, state government, multifamily residential, institutional
Amount	custom large business energy initiative program: technical service, financial services, and 50% of the project cost custom small business: up to 70% of project costs: remaining share financed by National Grid with a 0% interest loan: payback time of up to 24 months. fluorescent lighting: $15–$50/fixture LED exit fixtures: $10 LED: $15–$150 pulse start metal halide: $50 or $70 high-intensity discharge: $75 or $100 lighting sensors: $15–$60 compressed air: $180–$280/HP storage incentives: $2.75/gallon energy management system: $225–$275/point hotel occupancy sensors: $75/sensor refrigerated vending machine: $55 nonrefrigerated vending machine: $30 glass front refrigerated coolers: $75
Expiration Date	12/31/2011
Web Site	www.powerofaction.com/efficiency (National Grid 2011)

Source: NCSU 2011

Table 2-3. New York State Existing Facilities Program—from DSIRE Database24

State	New York
Incentive Type	State Rebate Program
Eligible Efficiency Technologies	equipment insulation, lighting, lighting controls/ sensors, chillers, furnaces, boilers, heat pumps, central air-conditioners, CHP (combined heat and power)/ cogeneration, steam-system upgrades, programmable thermostats, energy management systems/building controls, motors, motor variable-frequency drives, processing and manufacturing equipment, custom/others pending approval, commercial cooking equipment, commercial refrigeration equipment, data center equipment, food service equipment, interval meters
Applicable Sectors	commercial, industrial, nonprofit, schools, local government, state government, installer/contractor, federal government, agricultural, institutional
Amount	**prequalified measures**: varies **electric efficiency**: $0.12 per kWh (upstate), $0.16 per kWh (downstate) **natural gas efficiency**: $15/mmbtu (upstate), $20/mmbtu (downstate) **energy storage**: $300 per kw (upstate), $600 per kw (downstate) **demand response**: $100 per kW (upstate), $200 per kW (downstate); bonus incentives available **CHP**: $0.10 per kWh + $600 per kW (upstate) or $750 per kW (downstate) **industrial process efficiency (IPE)**: Same as electric and natural gas performance incentives **operational changes (as part of IPE)**: $.05/kWh (electric) and $6/MMBtu (gas) **monitoring-based commissioning**: $0.05/kWh **super-efficient chiller bonus**: $1400/kW (full load) or $1000/kW (NLPV)

Table 2-3. New York State Existing Facilities
Program—from DSIRE Database24 *(continued)*

Maximum Incentive	**prequalified measures (general)**: $30,000 (electric), $30,000 (gas) **prequalified measures (national fuel gas)**: $25,000 **electric efficiency and energy storage**: 50% of cost or $2 million **natural gas efficiency (general)**: 50% of cost or $200,000 **natural gas efficiency (national fuel gas)**: 50% of cost or $25,000 **demand response**: 75% of cost or $2 million (limit also applies to combined performance based efficiency and demand response measures) **CHP**: 50% of cost or $2 million **industrial process efficiency**: 50% of cost or $5 million **monitoring-based commissioning**: 50% or cost or $500,000
Eligible System Size	**performance-based incentives (except national fuel gas)**: $10,000 minimum cost (smaller projects may use pre-qualified incentives) **CHP**: 250 kW (whole system or size of addition)
Equipment Requirements	vary by measure
Funding Source	system benefits charge (SBC); energy efficiency portfolio standard (EEPS)
Program Budget	$47.9 million (electric); $8.6 million (gas)
Expiration Date	11/30/2011 (general); 06/30/2011 (CHP and demand response)
Web Site	www.nyserda.org/Programs/Existing_Facilities/default.html (NYSERDA 2011b)

Source: NCSU 2011

Utility rate structures can be very complicated and may be set up in such a way as to penalize energy use reductions (energy cost per unit of energy goes up as energy use is reduced). There are sometimes a wide variety of rates that a building could be on, and buildings are sometimes on the wrong rate for energy cost savings. Impacts of fuel cost adjustment clauses and potential impacts of changes in fuel costs passed on to the consumer should also be understood.

Before energy goals are set, a general understanding of potential impacts of energy reductions on actual costs of energy should be evaluated briefly. For building with kW demand charges, the energy savings may or may not impact the demand charge, which is typically set in the maximum 15-minute usage for the billing period. Also, the energy or demand savings may result in the building being placed on a different utility tariff.

APPLICATION TO SMALL BUILDINGS OR THOSE WITH LOW TO MODERATE ENERGY BILLS

Small-building managers or owners who read this book should benefit most from perusing the contents for recommendations that apply, checking a few of the Web sites, and then making contacts to find resources to help them. If the right resources can be found to proceed, then consult this book on specific issues as needed to help go through the entire process. Owners of small buildings also need to develop a simple system to track their energy use over time to verify and maintain efficiency improvements.

Energy costs must drive decisions about what comprises an energy efficiency program. For buildings or portfolios with low annual energy costs, building managers or owners will find difficulty in trying to apply resources to energy efficiency activities. However, an energy audit is appropriate to ensure all cost-effective measures have been addressed.

A national energy efficiency program for small buildings would benefit all stakeholders. Some states such as Wisconsin (Wisconsin Focus on Energy: www.focusonenergy.com [Focus on Energy 2011]) and New York (NYSERDA Energy Audit Service: www.nyserda.org/programs/energyaudit.asp [NYSERDA 2011a]) have programs that provide some level of energy efficiency services to smaller buildings. However, even more effective programs are needed to make it easier for smaller buildings to achieve higher levels of energy savings.

The Internet is a good source for building managers or owners to look for advice on how to address energy efficiency. Some sites provide advice for small businesses or small buildings, such as:

1. The DSIRE (Database of State Incentives for Renewable Energy) is a comprehensive source of information on state, local, utility, and federal incentives, policies, and programs that promote renewable energy and energy efficiency (www.dsireusa.org) (NCSU 2011).

2. The Climate and Energy Project is a source of information on alternate energy. Building owners are advised to implement cost-effective efficiency measures first, as the cheapest energy is usually the energy not consumed. However, one major retailer has spent and continues to spend money on photovoltaics to help with public relations and good will. Wind turbines are more difficult to site in urban and suburban settings. Small businesses should be careful about extensive commitment to expensive renewable energy technologies in general until buildings are efficient (www.climateandenergy.org) (CEP 2011).

3. The U.S. Small Business Administration Web site lists some easy energy-saving tips (www.sba.gov) (SBA 2011).

4. The California Energy Commission Consumer Energy Center Web presence has some advice for energy savings in summer seasons. (www.consumerenergycenter.org) (CEC 2011).

5. The ENERGY STAR® program has a small business program that can help orient one to all the needs for an energy efficiency program, and one can join to obtain free technical support, information, and awards eligibility (www.energystar.gov) (EPA 2011). There is also a guide document, *Putting Energy into Profits,* available for download (EPA 2007).

6. Finally, if one is ready for more complicated options and information, the Energy Crossroads Web site at Lawrence Berkeley National Laboratory can be checked (http://eetd.lbl.gov) (LBNL 2011).

Although more resources are available on the Internet, these provide more than most small-building managers or owners will ever want to consider.

HIRING RESOURCES VS. IN-HOUSE RESOURCES

Whether to conduct the energy efficiency process using in-house staff resources, outside consultants, or a combination of both is an important decision. Factors that are important to consider are staff experience and availability, expertise, cost, and anticipated scope of the project.

Using In-House Staff

Larger organizations may be able to assign responsibility for all activities to their own personnel, a situation which then requires some important internal process-related requirements:

- Provide a clear process
- Address fears of operators
- Address skepticism and barriers
- Offer training if needed
- Require accountability
- Require adequate time for human resources

A *clear process* requires that goals or task activities are not defined in terms that imply some type of improvement in what personnel are already doing, such as "better operation" or "better attention to thermostat control." Once guilt for not already doing the work is introduced, the process is likely to fail. Instead, goals and tasks must be more clearly defined, typically in terms that are outside of expected practice, such as in the following examples:

- Verify actual HVAC energy use during occupied and unoccupied periods
- Test controls changes to reduce the use of HVAC and lighting in unoccupied spaces
- Test chiller and boiler resequencing to verify reductions in total HVAC energy use

A clear process also requires that management not be dismissive of workers abilities, or think that "anyone can do that," as indeed, often a higher level of skill is required to handle the tasks in the energy efficiency improvement process.

Fears that changes will make current jobs more difficult, or that operational problems might result, must also be addressed. Similarly, initial feelings of skepticism about savings, or knowledge about barriers, should be sought out, and the ways the overall process will address uncertainty and barriers should be communicated to all involved in the process. Accountability means that progress will be measured in terms that are understandable to those involved, and their accomplishments reported.

Since higher levels of skills, or some skill improvements, are often needed for workers to accomplish some of the tasks, decisions must be made about whether to provide training first; hire outside experts to work alongside existing personnel to train them while working together; or just use the outside experts, with existing workers picking up knowledge as they are able.

Finally, one major issue with existing staff is that they may be already handling two or more jobs, and simply adding to the overload is not helpful. Reasonable estimates must be made about the level of work required, and if existing staff are used, they must be given adequate time to do the work.

Using Consultants

Managing an energy efficiency project using internal resources presents several considerations. First and foremost is the question: does in-house staff have the needed experience and expertise to identify, manage, and implement a successful energy efficiency project? For less complex equipment replacements (such as lighting, motor, and unitary HVAC equipment retrofits), in-house staff often have sufficient capability. However, for more complex projects such as energy management and control system installations, HVAC system modifications, and other projects of significant scope, it is often better to hire a consultant to help manage and oversee the project scope development and implementation.

Second, in-house staff may have a limited awareness of technology options and may be "too close to the trees to see the forest." Hiring an outside professional gives an outside perspective for identifying and recommending appropriate energy efficiency upgrades.

As succinctly summarized by the California Energy Commission, "there are advantages and disadvantages to using a consultant instead of your own staff" (2000):

Advantages. A consultant can:

- Confirm and verify that the projects are feasible
- Obtain and use the latest technical and cost information
- Use computerized building simulation models which can more accurately estimate project feasibility
- Identify technological problems before installation
- Free up your staff

Disadvantages. Contracting out does not mean no work for your staff. Your staff will still need to:

- Prepare the RFP (request for proposal) and the scope of work to select the consultant. Preparing an RFP is very time-consuming.
- Manage the consultant and review the energy audit.
- Resolve protests and conflicts from losing bidders and consultants.

While many customers are hesitant to hire a consultant due to the perceived added cost to the overall project, it's important to consider that using in-house resources takes staff away from other responsibilities and tasks, which has a cost and productivity loss.

To get the most out of an energy audit, the individuals preparing the scope of services and hiring the auditor must understand the scope of services and expectations up front. Important elements include:

- The work scope clearly defines the goals of the audit.
- The auditor understands the financial criteria of the owner. Most owners have an internal rate of return requirement, or minimum payback period in years for an energy efficiency measure.
- The description of the energy efficiency measures should contain enough information such that a trained practitioner can understand what is being proposed. The description should include an itemized cost estimate.
- Energy savings should be calculated using actual marginal energy costs, including, for example, kW demand charges and kWh charges as opposed to blended rates.

For small business, many utilities and state government energy agencies offer cost-shared energy audit and equipment installation programs. While these programs are typically available for larger customers as well, the limited scope of potential energy efficiency upgrades available in a small building make these programs an excellent fit. Customers can get the needed expertise to help them reduce energy use while improving building operations at a significantly reduced cost. For example, the New York State Energy Research and Development Authority offers free or heavily subsidized audits for commercial buildings with a peak demand of < 100 kW, cost-shared studies and retrocommissioning for large buildings, and incentives for cost-effective energy efficiency measures that save electricity and natural gas. All programs are maintained on the NYSERDA Web site (http://nyserda.ny.gov).

For large commercial, institutional, and municipal facilities, retaining an energy consultant is almost always the recommended approach.

Hiring a Consultant. Energy auditing is a specialized task. It is quite different from new building or new or replacement systems design. Those tasks require looking at the building and considering the conditions that might occur on the most extreme day(s) in a multi-year period, and sizing the system to handle that load. That is quite different from the forensic detective skills needed by an energy auditor. There are some engineers, even architects and other professionals, that do design and also have energy audit skills, but these do not inherently go hand-in-hand.

When hiring an energy auditor, one should look for a professional with specific energy auditing experience. This person need not only understand the technical aspects of all the energy systems in the building and how they work together in meeting the site's needs, but also possess the communication and investigative skills to extract the critical operations details from the staff and occupants.

Often, firms use teams of staff to conduct an audit. While such teams are certainly acceptable practice, you should assure that, unlike the more common practice of sending the lower level staff to do the fieldwork, you employ an audit firm that utilizes the most experienced personnel to conduct the on-site survey. Most members of the team should be able to do the computations back in the office. It is, however, the eye of the experienced auditor combined with his or her honed interviewing techniques, along with that individual's understanding of the energy systems, that will determine which energy efficiency measures (EEMs) should be taken (or at least considered) in that building. That, above and beyond all else, is *the* most critical factor in the value of an audit. Often, the best EEMs do not appear on a list of typical measures. They are opportunities caused by unique conditions in the building and will only be identified by an experienced auditor.

The auditor is a forensic detective who will be combining the clues from the data complied in the pre-site-visits stage with observation (visual or verbal responses to the on-site interviews with operators and occupants), and with the data that has been collected (from nameplates or with monitoring equipment) during the visits. Combining all of this information and using a very critical "experience filter" enables the auditor to determine the potential opportunities to reduce energy costs for the building.

The individual employed as an energy auditor should have all the skills described above. While an engineering degree or a P.E. (physical engineer degree) is certainly of value, end users should require a certification in energy and buildings systems analysis. The longest standing of these is the Certified Energy Manager (CEM)—Association of Energy Engineers (AEE). More recent ones (in order of length of existence) include the Certified Energy Auditor (CEA)—AEE, and Building Energy Assessment Professional (BEAP)—ASHRAE. Some of the certifying organizations keep public listings of certified professionals available on the Web for end users to verify current status. Additionally, potential auditors should be able to provide references for other audits that they have conducted on facilities of similar scope.

Beyond the skill set described, assure that the auditor hired investigates not only major capital improvements but the low-cost/no-cost OM&M measures as well. These OM&M measures are most often the opportunities that will offer both the highest return on investment and assist in the persistence of savings over time. Lastly, you should be looking for an auditor who will take the time to educate staff on the why and how of more efficient operations, and leave you with an audit report that is a plan for both short- and long-term actions to be taken to improve the building.

USING ENERGY SERVICE COMPANIES

In *Energy Efficiency Guide for Existing Commercial Buildings: The Business Case for Building Owners and Managers* (Landsberg 2009), the precursor to this technical guide, we briefly discuss the use of an energy services company (ESCO). An ESCO provides a turnkey project, including an initial energy audit to identify and prioritize energy conservation measures, design, implementation and financing. Repayment occurs over time as a share of the energy savings that accrue from the project. After a period of years, the project is turned over to the owner.

Advantages. Using an ESCO has the following advantages to the owner:

- Less involvement in implementation than with a standard project
- Reduced or no upfront costs
- Immediate use and benefit of the new equipment
- Potentially, the financing can occur off balance sheet

Disadvantages. Use of an ESCO has the following disadvantages:

- The ESCO is in business to make a profit and total costs will be higher than without the use of an ESCO.
- Since payment to the ESCO is based upon energy savings, the owner and the ESCO must agree on a methodology to determine those savings. Some efficiency measures require expensive measuring and verification (M&V) to prove savings. This includes adjustments for weather, occupancy, and changes in building operations not related to the energy efficiency measures. Alternately, the owner can save on the cost of M&V and take all or some of the risk that the savings are actually occurring. Monitoring and verification procedures are described in ASHRAE Guideline 14, *Measuring Energy and Demand Savings* (ASHRAE 2002), and in the International Performance Measurement and Verification Protocol (EVO 2007).

Under a guaranteed savings scenario, the ESCO would complete an energy audit of the building. If the project moves forward, the cost of the audit is rolled into the total project cost. If the building owner does not proceed with the project, a fee is typically paid for the audit. If the contract is based on shared savings, the contract must last for a sufficient period of time for the ESCO to recoup its investment plus a profit. Therefore, the project should include a mix of measures that results in a reasonable payback period, such that the contract duration will be five to ten years, or less. Conducting an independent audit first and using a third party commissioning agent provides assurance that the ESCO has thoroughly investigated energy efficiency opportunities and installed the measures in a proper manner.

The customers of most ESCO projects are government agencies or institutions. ESCOs can help speed the implementation process for these institutions. Government agencies often have a long wait to get funding for a project and the project can be completed and paid for by the time such funding is approved. K–12 schools can often implement projects without the need for voter approval. Institutions such as colleges, universities, hospitals, and nursing homes plan to use their facilities long-term and are potential candidates for an ESCO project. Generally, the customer requires a performance guarantee in return for the increased cost of an ESCO project. If such a guarantee is not wanted or needed, alternate, less expensive means of financing should be explored.

As a general example, consider a $500,000 ESCO project that generates $120,000 of savings per year. The total project cost includes ESCO fees associated with completing the initial audit and project development, design, implementation, program management, profit, etc. Assuming the $500,000 project cost is financed, some portion of the annual savings is used to pay the financing company. If the financing term is seven years at a 5% interest rate, the annual payment for debt repayment is $86,400. Of the remaining $33,600 in savings, some portion, typically a prenegotiated fee, is paid to the ESCO for services that may include equipment maintenance or service and annual savings reconciliation reporting. The remainder of the savings then remains with the customer. After the financing term is complete, the customer retains 100% of the annual savings thereafter.

Additional resources can be found at the National Association of Energy Services Companies (NAESCO) Web site: (www.naesco.org/resources/sites.htm) (NAESCO 2011).

COMMISSIONING

Commissioning is strongly recommended, and third-party commissioning should be mandatory for energy efficiency measures that cost more than $100,000. Commissioning assures that the measures are performing as intended and that the projected energy savings will be realized. Ongoing commissioning ensures that the building systems remain in tune and protects energy efficiency investments.

REGULATORY AND MARKET EFFECTS

Each electric and natural gas utility is unique. Rate tariffs are set in negotiation with the state's public utility commission. Energy cost can be reduced through careful review of the tariff and positioning the building accordingly. If the building is eligible for more than one tariff, analysis should be performed to determine which tariff is most economically beneficial.

Some utilities offer interruptible tariffs. These tariffs permit the utility to temporarily shut service supply at times of peak demand in return for a more favorable rate. If the building has emergency generators that can carry the building at their continuous duty rating, or dual-fuel boilers, interruptible rates may offer substantial cost savings.

Energy is deregulated in many jurisdictions and savings can sometimes be achieved by procuring energy from a supplier other than the distributing local utility. In the case of electricity and natural gas, the commodity will still be delivered by the local utility, but may be purchased from a variety of sources. Most investor-owned utilities offer a list of alternate suppliers on their Web site, or the state Public Utility Commission is a source for this information.

Commercial electricity (and sometimes steam or even gas) tariffs are typically divided into consumption charges and demand charges. The demand charge is based upon the peak 15-minute or 30-minute usage of the building in the billing period. The rationale for this charge is that the utility companies must build and maintain generation capacity to meet and exceed the highest demand they experience, even if it is only for one hour per year. Money can be saved by controlling when a building uses energy. Under some rate structures, demand can account for a very significant portion of the bill. Additionally, many electric utilities are switching to real-time pricing—the price of electricity changes every hour. With this change, it will be even more important to be able to control when a building uses energy.

Some electric utilities send pricing signals designed to lower high-demand peaks in the form of high peak-demand charges and low night-time electric rates. For customers served by those utilities, thermal storage systems or equipment scheduling strategies can produce substantial energy cost savings.

In many jurisdictions, it is possible to obtain lower electrical rates, or to be paid for electric capacity that can be turned off if there is a utility system demand approaching system peak that may result in brownouts or outages. In some cases, the system operator will pay the building owner for joining the program and agreeing to shed load when asked. Additional money is earned by the building owner if called upon to shed load. This will require rescheduling equipment, or running emergency generators when asked, to reduce the building electric load. The number of hours per year that a building must shed load to participate in the program is typically small, and ample notice is given to permit the building operator to take action. In some years, a building may never be called upon to shed load, yet still receive payments.

Measuring Energy Performance And Setting Goals 3

The basic methods of measuring energy performance and setting goals, as shown in Figure 3-1, are presented in this chapter. Once the types of performance goals are selected, existing building performance can be determined relative to those desired goals. For example, if the goal is to be at least 15% below the national "typical" energy and cost indexes for a specific building type, then the indexes calculated for the building are compared to the national "typical" values to see what level of reduction is needed to be 15% below. Alternatively, if the type of goal involves simply achieving a 15% savings relative to where the building is now, then the 15% improvement is calculated against the existing building index. Calculation of building indexes, as

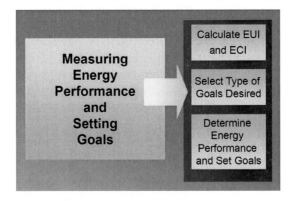

Figure 3-1. Measuring energy performance and setting goals.

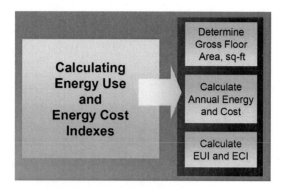

Figure 3-2. Calculating EUI and ECI.

shown in Figure 3-2, is covered in the next section. Since most building energy use scales with building size, gross floor area is used to normalize energy use per square foot such that buildings can be compared. Often building managers advance from simple goals to more complex goals over time.

CALCULATE ENERGY AND COST INDEXES

The first step in understanding energy usage is the determination of the energy utilization index (EUI) for a building. The EUI provides a simple value to track building energy usage over time and also provides a basis for comparing building energy use to other buildings and to standardized tables. The lower the EUI, the more efficient the building. Similar to EUI is the energy cost index (ECI). The ECI reflects annual energy cost, while the EUI reflects annual energy use. The gross floor area must be determined in order to calculate total annual energy use (EUI) and total annual energy cost (ECI).

Determination of Gross Floor Area

Both EUI and ECI are ratios that have gross floor area in the denominator. Gross floor area is typically determined based on the outside dimensions of a building, and is thus often easier to measure than values like net usable floor area or gross usable floor area. Gross floor area is important in that most published values of EUI data for different building types are based on gross floor area, and if comparison with published data is to be used, gross floor area is needed. *Gross floor area* is defined in ANSI/ASHRAE Standard 105-2007: *Standard Methods of Measuring, Expressing and Comparing Building Energy Performance* (ASHRAE 2007b), as:

> **"Gross floor area:** the sum of the floor areas of all the spaces within the building with no deductions for floor penetrations other than atria. It is measured from the exterior faces of exterior walls or from the centerline of walls separating buildings but it excludes covered walkways, open roofed-over areas, porches and similar spaces, pipe trenches, exterior terraces or steps, chimneys, roof overhangs, parking garages, surface parking, and similar features."

Calculate Annual Energy Use and Cost

Once gross floor area is known or measured, EUI requires calculation of total annual energy use. The EUI equals total annual energy use divided by gross floor area. EUI is typically expressed in thousands of Btu, or $kBtu/ft^2$. Table 3-1: Annual Energy And Cost Calculation shows the calculation of these two annual quantities for electricity, natural gas, or fuel oil as building fuels, and one value of "Other" is shown, but one could add several "Other" fuels if needed. Conversion multipliers for a wide range of fuels can be found at http://epminst.us (EPMI 2011).

The conversion factors in this file generate millions of Btu/yr, while thousands of Btu/yr are needed in Table 3-1, so if this calculator is used, the values generated must be multiplied by 1000 and then entered in the table under "Energy, (kBtu/yr)." The natural gas factors assume the utility bill energy use values include elevation corrections, needed at altitudes greater than 2500 ft above sea level (see ANSI/ASHRAE Standard 105-2007 [ASHRAE 2007b] if corrections are needed). Typically, only one entry is needed for natural gas or fuel oil, depending on the units used for billing, and no values need be entered if the fuel is not used in the building. Match the units used here to the units shown on the bills.

In the table, the annual total of Fuel Use should be the sum of 12 consecutive months of billing for fuel use (e.g., sum of kWh of electricity) divided by the total number of days in the 12 monthly bills, times 365 days in a typical year. This is necessary because utility billing cycles can have more or less days than a calendar month and the

Table 3-1. Total Annual Energy and Cost Calculation

Energy Type	Annual Fuel Use, A	Conversion Multiplier to kBtu, B	Energy (kBtu/yr), = A · B	Energy Cost, $/yr
Electricity, kWh/yr		3.412		
Natural gas, therms		100		
Natural gas, cubic feet (CF)		1.03		
Natural gas, CCF (or 100s CF)		103		
Natural gas, MCF (or 1000s CF)		1030		
# 2 Fuel oil, gallons		139		
# 1 Fuel oil or kerosene, gallons		135		
# 6 Fuel oil, gallons		154		
Purchased chilled-water		0.001		
Steam		0.001		
Other				
Total				

total for the year may not equal 365 days. Similarly, the Energy Cost should be the sum of 12 consecutive months of cost for the fuel ($), divided by the total number of days in the 12 months, times 365. Conversion to kBtu/yr for all fuels is accomplished by multiplying the amount of fuel used by the conversion multiplier:

$$\text{kBtu/yr} = \text{Annual Fuel Use } \mathbf{A} \cdot \text{Conversion Multiplier } \mathbf{B} \qquad (1)$$

Once all fuel use is known in kBtu/yr, the kBtu/yr can be added to the Total kBtu/yr. Similarly, the annual energy cost for each fuel should be added to arrive at the total energy cost for all fuels combined.

Energy Use and Cost Indexes

Once the building gross floor area is known and annual totals of all energy use and energy costs are calculated, the EUI and ECI can be calculated as below.

$$\text{EUI} = (\text{Total kBtu/yr}) / \text{Gross Floor Area} = \underline{\hspace{1cm}} \text{ kBtu/ft}^2 \cdot \text{yr} \quad (2)$$

$$\text{ECI} = (\text{Total energy \$/yr}) / \text{Gross Floor Area} = \$\underline{\hspace{1cm}} /\text{ft}^2 \cdot \text{yr} \quad (3)$$

Example EUI And ECI. Assume a building is 10,000 gross square feet in area. It uses 150,000 kWhs per year at a cost of $0.11 per kWh, or $16,500 per year. It also uses 8000 therms of natural gas per year at a cost of $1.25 per therm, or $10,000.

$$\text{EUI} = (150{,}000 \text{ kWh/yr} \times 3413 \text{ Btu/kWh}/1000 \text{ Btu/kBtu} +$$
$$8000 \text{ therms/yr} \times 100{,}000 \text{ Btu/therm}/1000\text{Btu/kBtu}) / 10{,}000 \text{ gsf}$$
$$= 131 \text{ kBtu/ft}^2 \cdot \text{yr}$$

$$\text{ECI} = (\$16{,}500 + \$10{,}000) / 10{,}000 \text{ gsf} = \$2.65/\text{ft}^2 \cdot \text{yr}$$

Energy Star Portfolio

Manager EUI calculation can be done using computer-based tools that will carry out much of the process described. The most prominent of the public domain tools is the U.S. Environmental Protection Agency's (EPA's) ENERGY STAR® Portfolio Manager. Portfolio Manager (PM) can be use to create an EUI for a building, although one must be careful to use the "Site Energy Use Intensity" value (kBtu/ft²·yr), as PM reports both *site* and *source* energy (EPA 2008a). For common building types PM also generates a comparison to similar buildings. PM can be used to evaluate a building for an Energy Star Plaque and is used by LEED—EB (Leadership in Energy and Environmental Design—Existing Buildings) as well (USGBC 2008). Go to www.usgbc.org for more information on LEED.

PM requires a certain level of effort and locks one into that system for future EUI information, so users must decide if their needs are matched better if they calculate EUI on their own, especially since the energy use values will be needed to estimate major end uses of energy. ECI is not calculated by PM. PM is useful for comparing buildings to its peers but is not necessary to simply determine an EUI.

Site and Source Energy

The terms *site energy* and *source energy* refer to differences in the conversion multipliers used to calculate kBtu/yr. Source energy includes energy used to generate, store, and transport the energy to a building, while site energy does not. The information presented in this book is based on site energy. Site energy values can cause difficulties in comparing one building to another building; for example, when electricity is generated on site. Energy performance goals for a building can still be handled without problems, as long as performance goals are set relative to the EUI and ECI of the building.

The ENERGY STAR PM tool reports site energy but uses source energy for setting energy performance levels. Those interested in the conversion factors for source energy used by the PM tool can examine *Performance Ratings: Methodology for Incorporating Source Energy Use* (EPA 2008b).

SELECT THE TYPE OF GOALS DESIRED

For most buildings, EUI and ECI can be used to set goals, track performance, and compare energy performance to other buildings. As mentioned in Chapter 2, if users need to consider other types of energy-derivative goals, then the energy calculations in Table 3-1 can also be used as the basis for calculating energy-derivative quantities. If other types of goals, such as certifications for energy, environmental, or sustainability programs are desired, the requirements for those certifications must be studied and applied.

The focus in this book is on simpler energy-based approaches, and the complications of additional certifications will not be addressed directly here.

DETERMINE ENERGY PERFORMANCE AND SET GOALS

Energy performance and performance goals will be presented in this book only in terms of EUI and ECI, although derivative values like carbon emissions should be almost identical, if not identical, to consider (i.e., linearly related with percentage changes identical). ECI values will have more variation than EUI if compared across wide geographic areas, but within a given utility service area, changes in ECI values for similar size and type properties should be reasonably comparable.

For certain types of buildings, it may be desirable to evaluate energy use against metrics other than per square foot. Examples of other metrics include energy use per occupied hotel room or energy use per hospital bed.

Goals can also be set relative to national certification programs, like the ENERGY STAR program, where data entered into Portfolio Manager are used to generate a program score or rating, and if the score is high enough, the building may be eligible for certification as an Energy Star building (EPA 2008a).

Simplest Performance Goal

The simplest and easiest performance goal to select is a fixed percentage lower EUI and ECI than the calculated EUI and ECI values. Few buildings can afford to invest enough to save 30% of all energy used, although some buildings that are highly inefficient can save 50% or more of their energy (reduce the EUI by 50% or more). Experience has shown that percentage improvements achieved for buildings can be categorized as shown in Table 3-2: Energy Savings Levels.

For most buildings, the simplest way to set a performance goal is to select from the categories of Low, Typical, and High in as shown in Table 3-2, and aim for savings (or EUI and ECI reductions) in the range chosen. High will be ambitious but attainable in many cases. Typical is more cautious and a good choice for an initial goal. Low is for those who are even more cautious and want to establish the process better before committing to more aggressive goals. Goals should be reassessed periodically as experience is gained in improving energy efficiency. This should occur at least every three years. It can be beneficial to mix long payback items with short payback items to bundle integrated projects. This is part of the LCC (life-cycle cost) approach and helps justify funding for projects that otherwise wouldn't have a reasonable payback. When balancing project financing, one shouldn't arbitrarily address all the "low-hanging fruit" first, although this is a good starting point for project planning.

The EUI and ECI calculated initially become the baseline performance values against which performance achievements and success in meeting goals are measured.

Table 3-2. Energy Savings Levels

Savings Level	Percent of EUI or ECI Saved
Low	5%–10%
Typical	10%–20%
High	20%–40%
Rare	40%–70%

Comparing to National EUI and ECI Data

For those who wish to compare their buildings to the national data in the U.S. Department of Energy (DOE)/Energy Information Administration (EIA)'s *Commercial Buildings Energy Consumption Survey* (CBECS) (EIA 2003) data base (accessible at www.eia.gov/emeu/cbecs), two tables of typical values are provided next. These baseline values are median values; half of all buildings are higher and half of all buildings are lower than the values in the table. In this way, one can see if a specific building EUI or ECI is higher or lower than the "middle-of-the-road" for the whole country. The ratings are scaled by building type and are adjusted for climate, operating hours, and other variables. It should be noted that the site and source EUIs in the ENERGY STAR data base are expressed in normalized terms and are adjusted for occupancy, plug/process loads and weather. Table 3-3: 2003 CBECS Typical National EUI Values provides the EUI values for buildings in the year 2003, and Table 3-4: 2003 CBECS Typical National ECI Values provides the similar "typical" ECI values. The 2003 CBECS was the last year data were available at the time of publication (EIA 2003). While national data are useful, some building types have significant variation by climate zone. Data by climate zone are available in Appendix A.

Table 3-3. 2003 CBECS Typical National EUI Values

Building Use Description	Typical Site Energy EUI $(kBtu/ft^2 \cdot yr)$
Education	
College/University (Campus-level)	CBECS does not have data on college campuses.
Elementary/Middle School	55
High School	78
Other Classroom Education	45
Preschool/Day Care	78
Food Sales	
Grocery Store/Food Market	178
Convenience store (without gas station)	263
Convenience store (with gas station)	230
Other Food Sales	73
Food Service	
Fast Food	419
Restaurant/Cafeteria	257
Other Food Service	147
Health Care	
Hospital/Inpatient Health	198
Nursing Home, Assisted Living	126
Clinic/Other Outpatient Health	72
Medical Office (nondiagnostic)	43
Medical Office (diagnostic)	44
Laboratory	266
Lodging	
Hotel	76
Motel or Inn	73

Table 3-3. 2003 CBECS Typical National
EUI Values *(continued)*

Building Use Description	Typical Site Energy EUI (kBtu/ft^2·yr)
Dormitory/Fraternity/Sorority	74
Other Lodging	71
Mall	
Strip Mall	94
Enclosed Mall	94
Office	
Administrative/Professional Office	67
Bank/Financial Institution	89
Government Office	77
Mixed-Use Office	78
Other Office	59
Other (all other types, which is a very wide range)	70
Public Assembly	
Entertainment/Culture	46
Library	94
Recreation	48
Social/Meeting	59
Other Public Assembly	42
Public Order and Safety	
Fire Station/Police Station	98
Courthouse	93
Religious Worship	38
Retail	
Retail Stores (not mall)	48
Other Retail	92
Vehicle Dealerships/Showrooms	82

**Table 3-3. 2003 CBECS Typical National
EUI Values *(continued)***

Building Use Description	Typical Site Energy EUI $(kBtu/ft^2 \cdot yr)$
Service	
Vehicle Repair/Service Shop	43
Vehicle Storage/Maintenance	28
Post Office/Postal Center	71
Repair Shop	48
Other Service	90
Storage/Shipping/Warehouse	
Self-Storage	7
Non-Refrigerated Warehouse	19
Distribution/Shipping Center	34
Refrigerated Warehouse	126
Vacant	11

Source: Calculated based on DOE/EIA 2003 CBECS microdata with malls (EIA 2003).
Notes:
1. Propane calculations based on methodology similar to ENERGY STAR® (EPA 2008), with some deletions based on uncertainties in propane use.
2. Buildings larger than one million square feet have inappropriate data and were deleted.
3. Except for vacant buildings, any building used less than nine months/yr was deleted.
4. Malls do not have the months of use variable coded, so malls were not included in the months of use criterion.

Table 3-4. 2003 CBECS Typical National ECI Values

Building Use Description	Typical ECI ($/yr/ft^2)
Education	
College/University (Campus-level)	CBECS does not have data on college campuses.
Elementary/Middle School	$1.07
High School	$1.01
Other Classroom Education	$0.94
Preschool/Day Care	$1.19
Food Sales	
Grocery Store/Food Market	$4.15
Convenience store (without gas station)	$5.26
Convenience store (with gas station)	$4.77
Other Food Sales	$1.80
Food Service	
Fast Food	$9.26
Restaurant/Cafeteria	$4.01
Other Food Service	$2.70
Health Care	
Hospital/Inpatient Health	$2.46
Nursing Home, Assisted Living	$1.61
Clinic/Other Outpatient Health	$1.53
Medical Office (nondiagnostic)	$1.06
Medical Office (diagnostic)	$1.18
Laboratory	$4.52
Lodging	
Hotel	$1.41
Motel or Inn	$1.22
Dormitory/Fraternity/Sorority	$0.87

Table 3-4. 2003 CBECS Typical National
ECI Values *(continued)*

Building Use Description	Typical ECI ($/yr/ft^2)
Other Lodging	$1.13
Mall	
Strip Mall	$1.96
Enclosed Mall	$1.70
Office	
Administrative/Professional Office	$1.37
Bank/Financial Institution	$2.00
Government Office	$1.41
Mixed-Use Office	$1.30
Other Office	$1.32
Other (all other types, which is a very wide range)	$1.08
Public Assembly	
Entertainment/Culture	$0.56
Library	$1.59
Recreation	$0.88
Social/Meeting	$0.71
Other Public Assembly	$0.83
Public Order and Safety	
Fire Station/Police Station	$1.41
Courthouse	$1.56
Religious Worship	$0.65
Retail	
Retail Stores (not mall)	$1.01
Other Retail	$2.07
Vehicle Dealerships/Showrooms	$1.54
Service	
Vehicle Repair/Service Shop	$0.82

Table 3-4. 2003 CBECS Typical National ECI Values *(continued)*

Building Use Description	Typical ECI ($/yr/ft^2)
Vehicle Storage/Maintenance	$0.47
Post Office/Postal Center	$1.16
Repair Shop	$0.78
Other Service	$1.58
Storage/Shipping/Warehouse	
Self-Storage	$0.20
Non-Refrigerated Warehouse	$0.35
Distribution/Shipping Center	$0.59
Refrigerated Warehouse	$1.53
Vacant	$0.33

Source: Calculated based on DOE/EIA 2003 CBECS microdata with malls (EIA 2003),

Notes:

1. Propane calculations based on methodology similar to ENERGY STAR® (EPA 2008), with some deletions based on uncertainties in propane use.

2. Buildings larger than one million square feet have inappropriate data and were deleted.

3. Except for vacant buildings, any building used less than nine months/yr was deleted.

4. Malls do not have the months of use variable coded, so malls were not included in the months of use criterion.

Understanding Energy Use And End-Use/ System Energy 4

In the previous section, calculation of building EUI was described. The total energy use of the building was compared to similar buildings to determine overall building energy efficiency. Buildings use energy in different ways. Major uses include:

- Heating
- Cooling
- Fans and pumps
- Refrigeration
- Lighting
- Office equipment
- Computers
- Other

Insulation and controls do not use energy directly but impact how much energy the building uses. A building may have an efficient cooling system but an inefficient lighting system. The pumping system may be efficient, but use too much energy because of poor controls. A large kitchen or computer facility may be the reason a building uses more energy than its peers. Understanding how a building uses energy, in addition to how much energy is used, will permit energy efficiency efforts to be focused on the systems that are the least efficient. This is the focus of Chapter 4.

CBECS END-USE ENERGY DATA

The amount of energy used for major energy systems must be understood in most cases to be able to estimate savings for many potential energy efficiency measures. In addition, any estimated savings for measures should be compared against total energy for the affected end use to make sure the estimated savings are not unreasonable. The DOE EIA 2003 CBECS (EIA 2003) data that were used to generate the national typical EUIs and ECIs also have estimated values for end uses of energy. These are shown in Figure 4-1. The analysis described here does not involve submetering, but rather involves estimates of lighting and equipment operating hours and seasonal energy allocations, relative to swing-season base loads for heating and cooling energy use—in other words, the use of simple spreadsheet models. End-use allocation can be precisely metered, but it is expensive to do so.

Heating, cooling, and ventilation are typically considered as one major end use: HVAC. The 2003 CBECS data show that HVAC accounts for approximately 51% (depending on climate and other

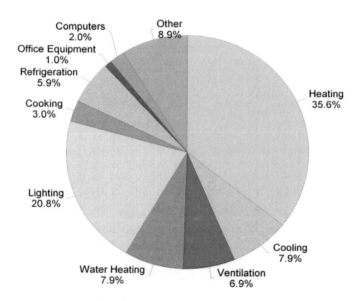

Figure 4-1. End-use percentages.

variables) of total energy used in commercial buildings in this country. Lighting accounts for 21%, and all other end uses account for the balance of 28%. This information was gathered from data table E-1a from the CBECS 2003 data, including malls (EIA 2003).

There are many types of commercial buildings, and the breakouts of major end-use energy differ greatly from one building type (use) to another. CBECS 2003 Table E-1a also has several end-use energy breakouts by different factors or characteristics (EIA 2003). Table 4-1, Building Category End-Use Percentages, shows the major end-use ratios of end-use energy divided by total energy for all buildings of that type across the whole country.

Most of the time, end-use breakouts should be estimated as percentages of a specific fuel use, and the means of estimating breakouts by major end use for individual fuel uses for specific buildings will be discussed in the next section. Since we want to estimate end uses by fuel, the CBECS data on end uses for specific fuels are of interest.

CBECS Electric End-Use Energy Percentages

Similar to the data used to develop Table 4-1, Building Category End-Use Percentages, other CBECS end-use data are available for electricity, natural gas and fuel oil, as well as district energy. Building category end-use percentages for electricity, natural gas, and fuel oil are presented in Tables 4-2, 4-3, and 4-4, respectively. These fuel-specific tables have fewer end uses with data that are not "uncertain," although the electricity table, presented in this section, has the same categories as the previous table.

CBECS Natural Gas and Fuel Oil End-Use Percentages

Table 4-3 presents results based on the CBECS end-use percentage tables for natural gas and Table 4-4 presents results for fuel oil (EIA 2003). Note that most values for fuel oil are uncertain and not able to be calculated, as there are fewer and fewer buildings using fuel oil, and statistical development of the data values becomes more difficult as the number of buildings lessens.

Table 4-1. Building Category End-Use Percentages

Principal Building Activity	End-Use Ratio Percentages (%)									
	Heating, %	Cooling, %	Ventilation, %	Water Heating, %	Lighting, %	Cooking, %	Refrigeration, %	Office Equipment, %	Computers, %	Other, %
Education	47.4	9.6	10.1	7.0	13.8	1.0	2.0	0.5	3.9	4.8
Food Sales	14.3	4.8	2.8	1.6	18.3	4.4	47.4	0.8	0.8	4.4
Food Service	16.6	6.8	5.6	15.7	9.8	24.6	16.4	0.5	0.5	3.7
Health Care	37.5	7.4	7.1	16.0	17.7	1.9	1.3	0.7	1.7	8.6
Inpatient	36.8	7.4	8.0	19.4	16.0	2.3	0.8	0.4	1.5	7.2
Outpatient	40.3	7.6	3.4	2.5	23.5	*	3.4	1.7	2.5	14.3
Lodging	22.2	4.9	2.7	31.4	24.3	3.1	2.4	*	1.2	7.1
Mercantile	26.3	10.8	6.7	5.6	30.2	2.5	4.8	0.8	1.1	11.3
Retail (other than mall)	33.5	7.8	5.0	1.6	34.8	0.9	6.9	0.9	1.3	7.5
Enclosed and Strip Malls	23.1	12.1	7.3	7.5	28.1	3.4	3.8	0.7	1.1	13.0
Office	35.3	9.6	5.6	2.1	24.8	0.4	3.1	2.8	6.5	9.7

Table 4-1. Building Category End-Use Percentages *(continued)*

End-Use Ratio Percentages (%)

Principal Building Activity	Heating, %	Cooling, %	Ventilation, %	Water Heating, %	Lighting, %	Cooking, %	Refrigeration, %	Office Equipment, %	Computers, %	Other, %
Public Assembly	53.0	10.3	17.0	1.1	7.3	0.8	2.4	*	*	7.0
Public Order and Safety	42.9	7.9	7.9	11.9	14.3	0.8	2.4	0.8	1.6	9.5
Religious Worship	60.1	6.7	3.1	1.8	10.4	1.8	3.7	*	0.6	11.7
Service	46.5	5.1	7.7	1.3	20.2	*	2.9	0.3	1.0	14.7
Warehouse and Storage	42.5	3.1	4.4	1.3	28.9	*	7.9	0.4	1.1	10.5
Other	48.3	6.3	3.8	1.4	20.6	*	3.5	*	1.7	11.5
Vacant	68.5	3.7	1.9	*	7.4	*	*	*	*	14.8

Source: Derived from CBECS 2003 data table E-1a, including malls (EIA 2003).
Note: Uncertain values are denoted by "*"

Table 4-2. Electric End-Use Percentages by Building Type

Principal Building Activity	End-Use Ratio Percentages (%)									
	Heating, %	Cooling, %	Ventilation, %	Water Heating, %	Lighting, %	Cooking, %	Refrigeration, %	Office Equipment, %	Computers, %	Other, %
Education	4.0	19.9	22.4	3.0	30.5	0.5	4.3	1.1	8.6	5.7
Food Sales	2.9	5.8	3.4	*	22.1	1.0	57.2	1.0	1.0	4.8
Food Service	4.6	12.9	11.1	4.6	19.4	6.0	32.3	0.9	0.9	6.9
Health Care	2.4	13.7	16.9	0.8	42.3	0.4	3.2	1.6	4.0	14.5
Inpatient	1.7	14.0	21.3	1.1	42.7	0.6	2.2	1.1	3.9	11.8
Outpatient	4.3	13.0	5.8	*	40.6	*	5.8	2.9	4.3	21.7
Lodging	6.0	10.2	6.0	5.1	52.8	0.9	5.1	*	2.6	10.2
Mercantile	7.9	14.9	9.3	5.2	42.0	0.3	6.7	1.1	1.5	11.3
Retail (other than mall)	2.8	11.8	7.6	0.9	52.6	*	10.4	1.4	1.9	10.4
Enclosed and Strip Malls	9.9	16.1	9.8	6.9	37.7	0.4	5.2	1.0	1.5	11.7

Table 4-2. Electric End-Use Percentages by Building Type *(continued)*

Principal Building Activity	End-Use Ratio Percentages (%)									
	Heating, %	Cooling, %	Ventilation, %	Water Heating, %	Lighting, %	Cooking, %	Refrigeration, %	Office Equipment, %	Computers, %	Other, %
Office	4.6	14.0	8.8	1.0	39.1	0.1	4.9	4.5	10.3	12.7
Public Assembly	3.0	21.0	37.7	*	16.2	*	5.4	*	1.8	13.8
Public Order and Safety	3.5	14.0	17.5	5.3	31.6	*	5.3	1.8	3.5	17.5
Religious Worship	4.8	17.7	8.1	*	27.4	*	9.7	*	1.6	29.0
Service	4.0	10.1	16.1	*	42.3	*	6.0	0.7	2.0	18.8
Warehouse and Storage	2.0	5.3	8.2	0.8	54.1	*	14.8	0.8	2.0	12.3
Other	1.5	12.0	8.3	*	44.4	*	7.5	*	3.8	16.5
Vacant	6.7	13.3	6.7	*	26.7	*	*	*	*	46.7

Source: Derived from CBECS 2003 data table E-1a, including malls (EIA 2003).
Note: Uncertain values are denoted by "*"

Table 4-3. Natural Gas End-Use Percentages by Building Type

Principal Building Activity	End-Use Ratio Percentages (%)			
	Heating, %	Water Heating, %	Cooking, %	Other, %
Education	77.2	13.8	1.9	7.1
Food Sales	69.2	5.1	20.5	*
Food Service	26.6	27.6	44.8	*
Health Care	56.0	30.5	4.1	9.5
Inpatient	50.5	34.8	4.4	10.3
Outpatient	89.5	7.9	*	*
Lodging	29.8	57.7	6.5	*
Mercantile	71.2	7.2	9.1	12.5
Retail (other than mall)	92.3	3.3	3.3	2.2
Enclosed and Strip Malls	60.5	9.3	12.8	18.0
Office	85.5	4.8	1.1	8.6
Public Assembly	90.2	2.0	2.9	*
Public Order and Safety	51.7	34.5	*	*
Religious Worship	93.9	2.4	3.7	*
Service	85.6	1.4	*	12.2
Warehouse and Storage	84.1	3.0	*	*
Other	82.8	2.3	*	13.8
Vacant	92.9	*	*	*

Source: Derived from CBECS 2003 data table E-7a, including malls (EIA 2003).
Note: uncertain values are denoted by "*"

Table 4-4. Fuel Oil End-Use Percentages by Building Type

Principal Building Activity	End-Use Ratio Percentages (%)			
	Heating, %	Water Heating, %	Cooking, %	Other, %
Education	95.7	4.3	*	*
Food Sales	*	*	*	*
Food Service	*	*	*	*
Health Care	54.5	18.2	*	18.2
Inpatient	55.6	22.2	*	22.2
Outpatient	*	*	*	*
Lodging	71.4	*	*	2.9
Mercantile	*	*	*	*
Retail (other than mall)	*	*	*	*
Enclosed and Strip Malls	*	*	*	*
Office	83.3	*	*	5.6
Public Assembly	96.6	*	*	*
Public Order and Safety	75.0	*	*	*
Religious Worship	100.0	*	*	*
Service	*	*	*	*
Warehouse and Storage	*	*	*	*
Other	*	*	*	*
Vacant	*	*	*	*

Source: Derived from CBECS 2003 data table E-9a, including malls (EIA 2003).
Note: Uncertain values are denoted by "*"

MAJOR SYSTEM/END-USE BREAKOUTS

Use the annual energy-use totals determined in Chapter 3 for each fuel type—kWh/yr for electricity and kBtu/yr for other fuels—to break energy use into important end uses. This is critical for setting priorities. Understanding how the building uses energy is also valuable in checking energy savings estimates, and possibly in some cases, as an input to savings calculations. Cost breakouts are not usually considered at this stage of the process. Use of a consultant is often required to achieve these energy breakouts reasonably, but building managers and owners can use in-house knowledge if possible.

The two major cases covered in this book for energy use breakouts are: all-electric building and electric/fuel building. An all-electric building has no other fuel supplied or used, except possibly minor amounts of natural gas or propane for small kitchen uses. Buildings with only minor uses of other fuel should be treated as all-electric for major end-use breakout purposes. Buildings that use significant amounts of fuel, such as natural gas, fuel oil, or propane, for space heating, domestic hot water, or other use, should be treated as an electric/fuel building. Other fuels, such as district energy or coal, can also be treated in this fashion, but the presentation here will not mention other fuels by name.

Electric/Fuel Building Breakouts

The basic objective is to understand how much of each fuel is used for each important end use of heating, cooling, lighting, and HVAC fans. For buildings such as grocery stores, refrigeration energy use is also typically important. For lodging buildings, inpatient health care buildings, and some others (see Table 4-1: Building Category End-Use Percentages in the previous section to judge), understanding hot-water energy use is important. Other end uses may or may not be important to estimate, depending on the percentage of energy used in that end use and on the priority of the energy efficiency measures. Iteration in setting priorities may be needed in some cases, starting with initial end-use estimates determined as quickly as possible, followed by initial setting of priorities, and then possibly followed by an iteration on refining end-use estimates and reexamining priorities.

Buildings with fan systems having static pressure control of any type should have fan electricity estimated separately from heating and cooling electricity. Buildings with smaller HVAC units that do not have static pressure controls of any type for HVAC fans should include the fan electricity use under heating or cooling. Any fan systems that run 24 hours per day and 7 days per week should always have an estimate made of annual electric use for that fan or group of fans.

Electricity use should be handled in terms of kWh/yr. Fuel use should be handled in kBtu/yr. If multiple fuels are used for the same end use, the values from the previous calculation of EUI should be summed to obtain total fuel use applicable to a particular end use. For example, if part of a building is heated using natural gas and part is heated using an oil-fired system, the sum of the energy for gas and oil should be used in calculating heating energy use for the whole building. Alternatively, if the space types are very different, such as the front of building being an office and the back being a warehouse, separate calculations can be made for each space type.

In order to make such end-use breakout calculations, one must understand for which systems the different fuels are used. As an Example, if natural gas is only used for heating, the calculation of the end-use breakout for heating is simply 100% of gas energy times the total annual kBtu/yr of natural gas use. Analysis of end-use breakouts is covered in the section following the next discussion of all-electric buildings.

All-Electric Building Breakouts

For all-electric buildings, the electricity use must be broken out to the major end uses. This type of calculation is difficult when heating and cooling are both provided by electricity. Most of the same concerns described previously, under Electric/Fuel Building Breakouts, still apply, except there is only one fuel to consider and all final values are in terms of kWh/yr.

Analysis of End-Use Breakouts

The objective, in simple terms, is to calculate or estimate the amounts of electricity use or fuel use for each of the end uses in one of the appropriate tables: Table 4-5, Electric/Fuel Breakout, for those types of buildings and Table 4-6, All-Electric Breakout, for all-electric buildings. If the building is primarily only one space type, then the "Space Type 2" values are not needed. If more than two space types or additional end uses are considered necessary, the basic concept can be expanded easily on a pad of paper or a spreadsheet.

Measurement can be used to provide better data on energy consumption by system or by end use.

Table 4-5. Electric/Fuel Breakout

End-Use Type	Annual Electric Use, kWh/yr	Annual Fuel Use, kBtu/yr
Space Type 1		
Heating		
Cooling		
Lighting		
Other (e.g., hot water, exhaust fans, other)		
Space Type 2		
Heating		
Cooling		
Lighting		
Other (e.g., hot water, exhaust fans, other)		

Table 4-6. All-Electric Breakout

End-Use Type	Annual Electric Use, kWh/yr
Space Type 1	
Heating	
Cooling	
Lighting	
Other (e.g., hot water, exhaust fans, other)	
Space Type 2	
Heating	
Cooling	
Lighting	
Other (e.g., hot water, exhaust fans, other)	

MEASUREMENT

Overall building energy use is quantified by the utility bills. However, it can be valuable to measure the performance of specific building systems. This is accomplished through submetering, or may be available from the building EMS. The types of measurements that should be made depend on the nature of the building system. The simplest measurement is of a parameter that is constant and that operates continually, or under a known schedule, such as a building exit sign. Efficiency improvements can be determined by an instantaneous measurement of the kW draw of the exit sign.

Slightly more difficult is a parameter that is constant when operating, but operates at an unknown schedule. Examples include constant-speed fans or pumps, or a lighting system controlled by an occupancy sensor. The power draw of these systems can be measured instantaneously. For the lighting system, footcandle readings can be taken when the lights are ON to verify lighting levels. Additionally, the operating schedule must be measured. Lighting loggers can be place in light fixtures to determine when the lights are operational. Motor loggers sense the magnetic field or vibration of the motor and record when the motor is operating. These measurements can be used to determine power use indirectly and as such are referred to as proxy measurements.

The most complicated parameters to measure are those with both varying schedule and varying usage. Examples include dimmable lighting systems, variable-frequency drives on pumps and fans, and chillers. In these instances, power must be measured. Sometimes it is easier to take one-time power measurements and use a proxy for interval data. For example, one-time measurements can be taken on a fan for a range of frequencies and the frequencies can be monitored. Power consumption can be calculated by correlating the frequency trends to the one-time power measurements.

Generally, data are gathered to support energy efficiency calculations or to determine how a system is operating. Begin by determining what parameters are needed and then decide the best way to measure them. In energy efficiency calculations, the equations should be examined to determine what parameters are unknown. Then select a method of measurement for those parameters. For building systems, decide what needs to be determined and then select parameters. For example, trending the position of the outdoor air damper on a fan can be used to determine whether it is closing when the building is not occupied and whether it opens completely during economizer operation.

Total building energy use is obtained from the utility bills. Operations of a specific system can be determined by point measurements. The most costly measurements in most buildings are separation of energy consumption by end use, as buildings are not typically wired for ease of measurement by end use. The ASHRAE publication *Performance Measurement Protocols (PMP) for Commercial Buildings* provides recommended measurement protocols at three levels of detail for energy, water, and indoor environmental quality (ASHRAE 2010e).

The types of data that can be gathered are described in the following sections.

Operational Characteristics

Operating Schedules. Operating schedules are readily available through the building controls, if the building has an energy management system. One of the best opportunities for energy savings is turning equipment and systems OFF when not needed. The operating schedules for all building systems should be reviewed to ensure they are correct. Examples of schedules-related data are as follows:

- Occupied/unoccupied hours in each controlled zone
- Warm-up and cooldown periods
- Unoccupied override conditions and timers
- Flow control resets (e.g., fan static pressure with VAV [variable-air-volume] box dampers)
- Equipment operating temperature resets (e.g., boiler with outdoor air)
- Cooling tower condenser-water resets

Set Points. Set points can be determined from the building controls. Set points should be reviewed to ensure they are set at the values intended by the building operator. Typical set points include the following:

- Space temperature
- Space humidity

- Space lighting levels
- Minimum outdoor airflow rates
- Carbon dioxide levels
- Economizer limits
- Boiler temperature
- Chiller temperature
- DHW (domestic hot water) storage and delivery temperatures
- Fan system flow control static pressure
- Water loop system flow control static pressures
- Heating and cooling system enable conditions

Operating Conditions

Loads. The efficiency of building equipment often varies with the load conditions under which the equipment operates. Unfortunately, loads can be difficult to measure accurately. For example, the output of a chiller is determined by the flow rate times the differential temperature entering and leaving the chiller. Since this differential might be 4°F–8°F, a small variation in accuracy between the two sensors can lead to a large error. The flow measurement in itself might need to be inferred (e.g., pressure drops, etc.) rather than directly measured. Some typical loads of interest-to-energy efficiency are:

- Heating and cooling
- Peak loads—electric, cooling, and heating
- System start-up loads
- Lighting power
- Fan and pumping power

Knowledge of building loads can be important in controlling or shifting loads to reduce peak electric demand.

Equipment Performance. The load and the part-load equipment efficiency determine the energy use of the system at any given time. Therefore, measurement of the operating efficiency of a system is only meaningful if the load is known. Typical efficiencies to be evaluated are:

- Combustion efficiency

- Cooling efficiency
- Energy recovery effectiveness

The information required to calculate these efficiencies may range from one-time measurements to multivariable data trending, depending upon the specific system being analyzed.

The data that are readily available in any building will be a function of the sophistication of the energy management systems (EMS) in the building. In some buildings, especially newer and larger buildings, extensive data will be available. Figure 4-2 (ASHRAE 2011d) shows the performance of three chillers over varying load ranges. In smaller buildings having unitary controls, there may be no recording capability at all. One should take advantage of available data. When data are not present, the value of collecting additional data will depend upon the energy savings potential for the system and the degree of uncertainty with regard to savings estimates.

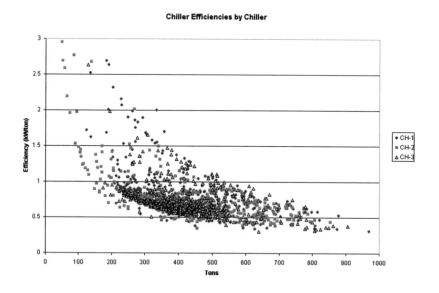

Figure 4-2. Time series data can show revealing trends.

Measurement Options

Spot Measurements. Spot measurements are used to determine the values of steady-state systems that are either ON or OFF. The connected load of a lighting circuit with all the lights on or airflow in a constant volume fan system are represented well by a single sample measurement. Typical operating values can be made by spot measurements. Figure 4-3 shows spot measurements of space temperature and humidity and lighting levels. Some data are available from the building EMS system.

Interval Measurements. Trend data are useful in verifying operations within schedules or relationships between primary and secondary variables (e.g., outdoor air temperature vs. heating hot-water temperature).

Duration. Measurements may be taken over a short or long time frame, a period of hours, days, weeks, months, or seasons. It is often useful to take data during both weekdays and weekends or holidays, as building operations are typically different.

Interval. Measurement intervals of 15, 30, or 60 minutes are common in building controls systems because these interval lengths give a good indication of how the building follows operating schedules. Longer intervals provide a trade-off with less data storage at the expense of providing less detail. A shorter interval (e.g., 1 minute) may be justified

(a) **(b)**

Figure 4-3. Example spot measurements: (a) space temperature and (b) humidity and light levels.

(Photograph courtesy L&S Energy Services, Inc.)

for a short duration in order to assess dynamics of the control system tuning.

Building control systems can often be used to trend monitored parameters. Observing the performance of a system over a week or two is an excellent method of determining how the system is performing, as shown in Figure 4-4 (ASHRAE 2011d). The data can be stored and either graphed or exported to Excel data files. Many EMS systems lack sufficient storage for the trending data, but adding storage is inexpensive.

Annual, monthly, and daily intervals are useful in looking at total energy use. They provide a high level view of operations useful for benchmarking and comparisons to historical data. Daily data are particularly useful in developing weather correlations, where the daily total electricity or gas use are correlated to the daily average outdoor air temperature to heating and cooling load lines. To some extent, monthly data can be weather normalized using heating and cooling degree days to determine the portion of energy use that is weather dependent. Annual totals provide a basis for historical comparison.

Data Collection Methods or Options

The following are potential data sources in a building.

Building Control Systems. Most building control systems offer some level of data trending. Data may be archived in a server for years for every point, or be limited to the last 24 hours of a few predetermined points. The ability to quickly assess the availability and scope of data are good skills for an auditor to develop. The building operator or the controls specialist are key sources of information. Their level of familiarity with the building control system features will provide the basis for determining data and might provide insight into their level of effort in ongoing evaluation of systems operations through the control system.

Some systems might be underutilized with few defined trends. It might be worth a small investment in time to learn about setting up trends on the particular system with the building operator.

Some systems store overwhelming amounts of data. Knowing the key operating parameters for the main systems being reviewed is vital in not having to deal with large amounts of data. Most systems can present a listing or graphic representation of data; however, it is often useful to capture the data for further off-line analysis in a spreadsheet

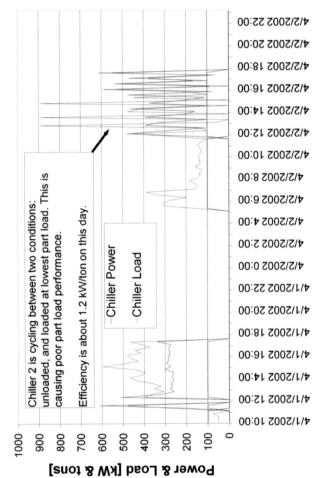

Figure 4-4. Example short-term interval monitoring.

to develop relationships among multiple points. Most systems offer data export features to common spreadsheet or text formats. Data in varying time frames and formats can be aggregated and analyzed by publicly available tools such as the Universal Translator (www.utonline.org).

Graphical representations of the systems can be useful. But it is important to verify that the graphics are representative of the actual system. Many systems have standard graphics that generally apply to a particular site, but sometime omit specific details. The graphics can provide an easy means to capture snapshots of an entire system's operation. Often screen captures or hard copy printouts are the only means to capture these data.

Stand-Alone Data Loggers. Data loggers are available as single-measurement, battery-operated devices or multimeasurement configurable devices. When no trend data are available from a building control system, these stand-alone loggers serve well to quantify operations. Runtime, status, temperature, humidity, and electric current are most common, but loggers exist for capturing data from any sensor with an output signal. Data intervals are often user configurable and duration is only limited by the amount of onboard memory and the precision of the data. Data loggers offer an advantage to building assessors in that their calibration can be maintained, whereas when trend data are used, the calibration must be established anew for each building site. In Figure 4-5, a battery-operated data logger is used to make a short-term power measurement on a circuit.

Handheld Meters. Handheld instruments with a known calibration condition provide both a good verification means of existing data and a source for new data. Temperature, humidity, carbon dioxide, airflow, lighting level, sound, amperage, and power are all readily measurable with handheld equipment. In Figure 4-6, the pressure in a hydraulic loop is being checked. Accessibility and the level of intrusiveness can limit the ability to use handheld meters. In Figure 4-7, a measurement bag and a watch are used to determine the flow rate of a sink faucet.

Existing Gages. Like building control system sensors, existing gages provide the same data used by the building operators. The accuracy of the gages both in measurement and in proper placement to be representative is vital. Using faulty data to assess operation will not

Figure 4-5. Installing stand-alone power logging equipment.
(Photograph courtesy L&S Energy Services, Inc.)

Figure 4-6. Measuring hydraulic pressure using a handheld meter.
(Photograph courtesy L&S Energy Services, Inc.)

Figure 4-7. Measuring water flow rate using a simple calibrated bag and timer.
(Photograph courtesy L&S Energy Services, Inc.)

provide any new insights to the systems. The auditor should consider the need to verify gages through comparison with calibrated handheld equipment and a verification of the sensor placement based on the importance of the data. For instance, understanding if the hot-water temperature gage is representative of the boiler temperature or the delivery temperature in a system that has both primary and secondary loop is important to clarify in documenting the system operation.

Accuracy

Erroneous data will lead to erroneous results. Accuracy must be considered for the individual measurement as well as the application of the measurement to be representative of the intended information. An independent comparison to a calibrated measuring device provides a high level of verification. In Figure 4-8, the accuracy and cycling of a thermostat are being determined. In this case, accuracy is important. In Figure 4-9, the operation of a DHW control is being verified. This is an ON/OFF control and absolute accuracy is of less concern. Sometimes it is possible to make relative comparisons of sensors measuring the same values when systems have no loads or energy inputs (e.g., supply and return water temperature across a chilled-water system that has chillers disabled). One should assess the needed level of accuracy for a specific device. A discharge air sensor from a variable-air-volume box as its as its primary function is to verify the operation of reheat when reheat is called for. Conversely, the accuracy of a building static pressure sensor used to control an exhaust fan speed would directly affect building energy use. Control system sensors that function in closed control loops are calibrated and provide feedback. Absolute accuracy is not as important as repeatability. If supply duct pressure is maintained by varying the fan speed, the absolute value of the pressure is not important if the duct pressure setpoint is reset based on air terminal damper positions. However, if the variable-speed drive (VSD) is set up to maintain pressure at a fixed absolute valve, a sensor might drive the system to operate at a needlessly higher pressure using more fan energy. In that instance, the accuracy of the sensor is important to overall system efficiency.

The more sophisticated the measurements, the more costly and complex the monitoring and verification will be, as shown in Figure 4-10. The cost of M&V can be minimized by reviewing calculations to isolate

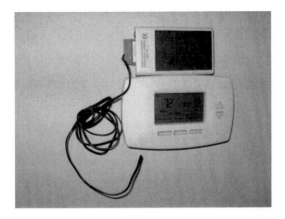

Figure 4-8. Verifying thermostat accuracy.
(Photograph courtesy L&S Energy Services, Inc.)

Figure 4-9. DHW control valve testing.
(Photograph courtesy EMRA.)

key parameters that need to be verified to determine the energy savings. M&V can then be focused on verifying the accuracy of those key parameters.

Example. This simplified example illustrates the basic concepts, without using extensive analysis.

The example building consists of a one-story 8000 ft^2 office section, with a one-story 20,000 ft^2 warehouse in back. "Space Type 1" will be considered the office, and "Space Type 2" will be the warehouse. Both spaces are heated with gas, but there are separate electric and gas meters on each space. The warehouse is not cooled and has no hot-water system. The warehouse has skylighting roof panels and is mostly used during daylight hours.

The office space uses 70,000 kWh/yr of electricity (as a reference, approximately $6000–$12,000/yr, depending on location) and 320,000 kBtu/yr of natural gas (approximately $3000–$5000/yr, depending on location). The warehouse space uses 40,000 kWh/yr of electricity (approximately $3500–$7000/yr, depending on location) and 360,000 kBtu/yr of natural gas (approximately $3000–$5000/yr, depending on location).

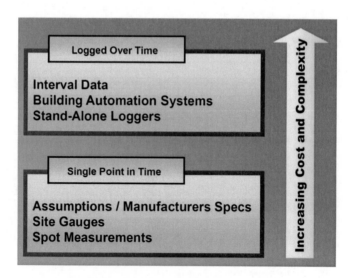

Figure 4-10. Cost and complexity for measurement strategies.

The building owner, his accounting person, and one technically oriented member of the company decide to make their best estimate of the energy breakout and come up with the following results. The lighting energy is estimated by counting light fixtures and knowing the approximate wattage of each. Lighting systems in both spaces are estimated to be ON about 3000 hr/yr. Lighting in the office space is estimated at 70 W per fixture, at 80 fixtures functioning, plus 2500 W of other lighting. Total lighting power is calculated as 70 × 80 + 2500 = 8100 W, or 8.1 kW. Lighting electricity use in the office is 8.1 × 3000 = 24,300 kWh/yr. This is 35% of total electric use, which is 4% less than shown in Table 4-7.

Based on the increase in electricity use in the months of June–September, it is estimated that the cooling electricity use is 26,000 kWh/yr. They know heating electricity is only for fans, and they decide to roughly estimate this use as 5000 kWh/yr for heating. The balance remaining is assigned to the "Other" category, at 14,700 kWh/yr. A similar analysis leads to the breakout for natural gas in the office and both electricity and gas in the warehouse.

Table 4-7. Example: Electric/Fuel Breakout

End-Use Type	Annual Electric Use, kWh/yr	Annual Fuel Use, kBtu/yr
Space Type 1		
Heating	5000	312,000
Cooling	26,000	0
Lighting	24,300	0
Other	14,700	8000
Space Type 2		
Heating	2000	360,000
Cooling	0	0
Lighting	37,000	0
Other	1000	0

MEASUREMENT RESOURCES

The following are resources for measurement.

ASHRAE Guideline 14-2002, Measurement of Energy and Demand Savings. This guideline was developed to provide methods for reliably measuring the energy demand and water savings of energy efficiency projects (ASHRAE 2002).

ASHRAE Handbook—Fundamentals. Chapter 36—*Measurement and Instruments* describes characteristics and uses of instruments used to measure building performance (ASHRAE 2009c).

ASHRAE Standard 105-2007, Standard Methods of Measuring, Expressing and Comparing Building Energy Performance. This Standard provides guidance in assessing building EUIs (ASHRAE 2007b).

International Performance Measurement and Verification Protocol. IPMVP provides best practice techniques for verifying results of energy efficiency, water efficiency, and renewable energy projects (EVO 2007).

Federal Energy Management Program M&V Guidelines: Measurement and Verification for Federal Energy Projects. FEMP M&V Guidelines can be found at the DOE Web site (EERE 2008b). The document provides guidelines and methods for measuring and verifying energy, water, and cost savings associated with federal energy savings performance contracts (ESPCs).

PRIORITIZING POTENTIAL ENERGY MEASURES

Developing the basic understanding of where energy is used in a building is a tremendous step toward understanding how to prioritize possible energy improvements. Prioritization will depend partly on how much energy reduction is desired to meet the energy goals selected. If the energy goal is simple, like reducing total energy use 15% from existing, prioritization will be affected primarily by which major end uses are the largest. It is important to remember that energy efficiency measures include more than capital cost measures. Items such as operational changes, schedule modifications, and efficiency tune-ups can be low-cost but result in significant energy savings.

In an example, an owner was interested in saving 10%–15% of total electricity use, and had vague expectations that the savings would occur in the office portion of the building. After looking at the energy-use breakout, the team decided they could reduce the warehouse lighting energy about 90%, which would be 30% of total electricity use of 110,000 kWh/yr (about 33,000 kWh/yr).

The existing warehouse lighting fixtures—typical metal halide— were noisy, too bright, and could not be turned OFF without causing restart issues (they were turned ON early each day to be working well by start of business). Light was needed mostly in a few key areas. Warehouse layout allowed a new approach, but not all warehouse layouts might be as flexible. New fluorescent fixtures that could be turned ON and OFF during the day were installed in the key areas, and the metal halide lamps were left OFF most of the time. The metal halide fixtures were run a few hours per year, when daylighting was not adequate in some portion of the warehouse, and additional light was needed. Expenditures of about $3500 saved about $4000/yr in electric costs.

One caution should be given about early success. Simply because some important savings are achieved right away does not mean more savings are not possible. The entire process can be started again, following initial efforts, as shown in the energy efficiency process flow diagram shown in Figure 4-11.

The importance of iteration on priorities is shown by the green arrow from the bottom to the top on "Reassess Process and Goals." Once additional goals have been achieved, further goal setting should be considered.

Priorities are set in the third step of the process, following end-use and system breakout analyses. The next step on energy and cost assessment may indicate a need to reassess priorities as this is a process of continuous reevaluation. The savings assessment will indicate the improvements that can be considered under the priorities as set, but some adjustment in priorities may allow additional efficiency measures to be considered. Also, energy savings should be implemented under an overall building plan. Just focusing on the "low hanging fruit" might not be the best strategy in the long run.

The overall building plan includes consideration of items such as the life cycle of building equipment and new building tenants. New technologies may also be a reason to reevaluate the plan. More sophisticated lighting controls and the development of LED (light-emitting diode) lamps are both progressing at a rapid rate and new products will shortly be available that should be considered for implementation.

END-USE ENERGY COSTS

In order to calculate cost savings for energy efficiency measures, knowing the total annual energy costs by major end use is needed. With the completion of the breakout of energy by major end uses, the annual energy-use allocation methods were considered, evaluated, and applied. In the same fashion, total annual energy costs by either electric/fuel or all-electric should also be allocated for use in some of the simplified costs savings calculation methods to be presented. For electric/fuel buildings,

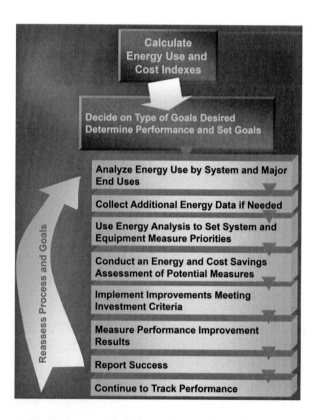

Figure 4-11. Energy efficiency process flow diagram.

the total costs for each end use for electricity and fuel, as shown in Table 4-8, should be added to "Total Annual Cost" for each end use.

As with the end uses of energy, if additional end uses or more space types are needed, they can be added.

In the last section, end-use electric energy consumption for a building that was part office and part warehouse was estimated. The energy cost by end use for an all-electric building can be estimated using Table 4-9. Assuming a cost of $0.12 per kWh and $1.25 per therm of natural gas, the annual energy use and cost per end use of a two-fuel building is shown in Tables 4-10 and 4-11, respectively. Note that one therm contains 100,000 Btus, so the cost per kBtu is $0.0125.

Table 4-8. Electric/Fuel Cost Breakout

End-Use Type	Annual Electric Cost, $/yr	Annual Fuel Cost, $/yr	Total Annual Cost, $/yr
Space Type 1			
Heating	$	$	$
Cooling	$	$	$
Lighting	$	$	$
Other (e.g., hot water, exhaust fans, other)	$	$	$
Space Type 2			
Heating	$	$	$
Cooling	$	$	$
Lighting	$	$	$
Other (e.g., hot water, exhaust fans, other)	$	$	$

Table 4-9. All-Electric Cost Breakout

End-Use Type	Annual Electric Cost, $/yr
Space Type 1	
Heating	$
Cooling	$
Lighting	$
Other (e.g., hot water, exhaust fans, other)	$
Space Type 2	
Heating	$
Cooling	$
Lighting	$
Other (e.g., hot water, exhaust fans, other)	$

Table 4-10. Example: Electric/Fuel Breakout

End-Use Type	Annual Electric Use, kWh/yr	Annual Fuel Use, kBtu/yr
Space Type 1		
Heating	5000	312,000
Cooling	26,000	0
Lighting	24,300	0
Other	14,700	8000
Space Type 2		
Heating	2000	360,000
Cooling	0	0
Lighting	37,000	0
Other	1000	0

Table 4-11. Example: Electric/Fuel Cost Breakout

End-Use Type	Annual Electric Cost, $/yr	Annual Fuel Cost, $/yr
Space Type 1		
Heating	$600	$3900
Cooling	$3120	0
Lighting	$2916	0
Other	$1764	$100
Space Type 2		
Heating	$240	$4500
Cooling	$0	0
Lighting	$4440	0
Other	$120	0

Selecting Energy Efficiency Measures 5

Chapter 5 is a description of energy efficiency and cost measures that are typically found in buildings. They are representative of the types of measures that should be considered in any building, but should not be the only measures to be considered. Often, significant savings can be realized through one or more measures that are unique to a particular building and will only be discovered through a detailed energy audit by an experienced energy auditor. Also, building systems are continually being improved and the latest proven technology should be considered when upgrading building systems.

The following lists and figures present a logical order for reviewing and selecting energy efficiency measures in a building. Measures are discussed individually in this section. However, once selected, the measures should become part of the energy efficiency plan for the building. An integrated approach that encompasses all cost-effective measures is the best way to proceed in the long run. Conditions in a particular building may result in reordering, but this process is appropriate for most buildings. The process overview in Figure 5-1 shows the flow of the decisions, and the measures topic list presents all the measure areas to be considered.

A typical order for undertaking energy efficiency improvements is as follows:

1. Operational changes (controls and maintenance, building tuneup)
2. Lighting and lighting controls
3. HVAC controls, maintenance, variable-speed motors, etc.
4. HVAC equipment replacement
5. Envelope measures

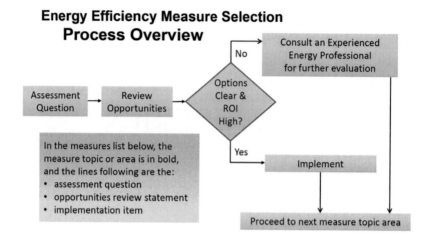

Figure 5-1. Measure selection.

This order may change, given specific building conditions and the life cycle of building systems. For example, the need for roof replacement may dictate implementation of roof insulation at an earlier point in the process than Step 5. Also, envelope improvements can reduce the loads on buildings systems and there may be good economic reasons to implement envelope improvements to reduce the size and cost of mechanical systems.

Considerations for Small Buildings

For small buildings, focus on the measures that are appropriate for the building in question. These include packaged and rooftop systems, lighting, thermostats, and operating schedules.

Selecting Energy Efficiency Measures

The energy efficiency process can be started by in-house staff. How far the process can be taken without assistance will depend upon the capabilities of building staff. Staff who are familiar with the building can review the EEM descriptions in Chapter 5. Some of these should be obvious high return on investment opportunities and can be implemented. Others will need further analysis. The following category list represents a logical approach for assessing the various systems in the building. Start with the first topic and move through the list, assessing conditions, identifying opportunities, and deciding whether they can be implemented immediately or require further study. Once all of the topics have been assessed, there will be a list of EEMs that can be implemented and a list that needs further work.

Some of the measures needing analysis can be evaluated in-house if the building staff has the technical capability to do so. Likely, an experienced energy analysis professional will be required to evaluate some of the measures. When they perform a survey, they are likely to identify additional measures that were not identified by building staff. The building owner/operator can then meet with the energy efficiency professional and finalize a list of measures to be evaluated, based upon the financial criteria under which the building is operated.

Utility/Fuel Supply: Is building taking advantage of supply and billing rate control/management issues? If not, review *Utility/Fuel Supply* opportunities. Implement clear and high ROI options and set other aside for further evaluation.

Operations: Have building operations been optimized? If not, review *Operations* opportunities—operating schedules should be emphasized. Implement clear and high ROI options and set other aside for further evaluation.

Building Envelope: Is the building shell tight and comfortable? If not, review *Building Envelope* opportunities. Implement clear and high ROI options and set other aside for further evaluation.

HVAC Equipment: Are the mechanical systems the most efficient? If not, review *HVAC Systems* opportunities. Implement clear and high ROI options and set other aside for further evaluation.

HVAC Distribution Systems: Do the heating and cooling systems deliver even comfort (at the lowest cost)? If not, review *HVAC Distribution Systems* opportunities. Implement clear and high ROI options and set other aside for further evaluation.

Water Heating: Is the hot-water system operating at optimal efficiency? If not, review *Water Heating* opportunities. Implement clear and high ROI options and set other aside for further evaluation.

Lighting: Have the building's lights and lighting controls been upgraded in the past two years? If not, review *Lighting* opportunities.

Implement clear and high ROI options and set other aside for further evaluation.

Motors: Have energy-efficient motors been utilized throughout the building? If not, review *Motors* opportunities. Implement clear and high ROI options and set other aside for further evaluation.

Controls: Are the existing controls systems being fully/optimally utilized? And/or has an automated Energy Management System been installed? If not, review *Controls* opportunities. Implement clear and high ROI options and set other aside for further evaluation.

Overall Performance: Does the building achieve a very low EUI? If not, hire an experienced energy professional to conduct a study or audit. Implement clear and high ROI options and set other aside for further evaluation by an experienced energy efficiency professional.

There are variations in savings opportunities by building system and building type. In some cases, important cost savings can result from switching utility rate schedules or switching fuels, although energy savings are often small or slightly negative for some fuel switching changes. Utility/fuel supply measures will only be covered briefly in this book, as they are not applicable in many buildings, and they also represent some complications for analysis. Fuel switching is typically done to save on energy costs or reduce emissions rather than to save energy.

Buildings having the greatest energy savings potential usually have energy systems operating many more hours per year than needed to serve building needs properly. In these cases, some means must be found to cause or allow the energy systems to operate less (reduce operating hours). Reducing energy system operating hours is an efficiency measure that typically offers the most energy savings potential in commercial buildings. Another simple means of saving energy in commercial buildings is by changing temperature setpoints to cause the heating and cooling systems to use less energy during unoccupied periods.

The energy efficiency measures most often presented in books or manuals are typically called retrofit measures, where some part of a building is changed or replaced, such as replacing incandescent light bulbs with compact fluorescent bulbs, or tuning up a boiler to increase efficiency. These retrofit measures will be presented under the major categories of: Utilities/Operational, Lighting, Controls, Motors, HVAC

Distribution Systems, HVAC Systems, Water Heating, and Building Envelope.

Many building systems such as chillers, boilers, pumps, and fans operate over a range of performance as the heating and cooling load on the building changes. Efficiency improvements must consider performance over the operating range of the equipment. For example, a boiler may be highly efficient at 100% load but only operate at that load for a few hours per year. Efficient performance at the range of conditions where the equipment actually operates is what should be considered.

UTILITY/FUEL SUPPLY

Change Utility Rates/Fuel Supply

With the advent of deregulation, electricity and natural gas can be purchased from multiple suppliers and delivered through the local utility. Both the local utility and the public utility commission will have lists of qualified suppliers. It is a simple matter to contact several suppliers, supply a year's worth of utility bills and obtain a price quote. Energy cost savings of 5%–10% are typical. Larger buildings can command larger discounts. Once energy is being purchased from another supplier, it is important to periodically (e.g., every two years) obtain new price quotes to ensure rates are still competitive. Some suppliers will print the cost savings as compared to the local utility on the monthly invoice. The building will still get a small invoice from the local utility for delivering the energy over their pipes or wires.

Switch Fuels

Many heating and central plant systems have the ability to operate on more than one fuel type. For instance, many boilers plants are designed and installed with the ability to operating on at least two fuels (e.g., natural gas or oil). Having the ability to choose between fuels enables the operator to select the least expensive fuel choice. For instance, if natural gas has a cost of $1.00 per therm ($10 per MMBtu) and #2 Oil has a cost of $2.50 per gallon ($18 per MMBtu) then the most economical fuel choice would be natural gas at nearly half the cost. However, if natural gas were to increase to $1.50 per therm ($15 per MMBtu) and #2 Oil dropped to $2.00 per gallon ($14.40 per MMBtu) then #2 Oil would be less expensive to use.

Other considerations would include carbon emissions, as natural gas has a lower carbon emission than fuel oils. The carbon emissions of natural gas are roughly 117 lb carbon dioxide (CO_2) per MMBtu and that of fuel oil (#2, #4, #6) is roughly 161 lb CO_2 per MMBtu. So to the extent that reducing carbon footprint is an issue, selection of natural gas versus fuel oil becomes important.

Reduce Peak Power Demand

Some utility/fuel supplier rate structures may have *demand* components or have demand-based rate blocks. With these types of structures, customers are charged not only for the amount of energy they use, such as kWh for electricity, but also for the generation capacity the utility is required to carry to meet the building's maximum load. Electric power, or demand, is measured in kW. Total energy used over a month or billing period is paid for as energy, while demand (or peak demand) is the highest amount of energy used in a short period of time, such as an hour or quarter-hour, during that billing period. Demand costs can be up to half or more of total charges for a fuel, so making changes to a building to reduce demand charges can be important for those utility or fuel costs impacted by demand. Some utility rates have a ratchet clause. A percentage of the peak 15-minute demand is used to set the minimum monthly demand charge for the next eleven or twelve billing periods.

The impact of demand on utility or fuel cost should be understood. If demand or demand-influenced factors are impacting costs, means of reducing those cost impacts should be evaluated. One important means of checking demand influence is to calculate the load factor (LF) of a fuel for a billing period:

$$\text{Electric Monthly LF} = \text{Monthly kWh} / (\text{"Demand" Monthly kW} \times \text{Total Billing Hours for the Month}) \qquad (1)$$

Smaller LFs (<0.15) are indicators of spiky profiles, which may indicate that there are high, short-lived peaks in demand by the building equipment. Office buildings often have electric LFs in the range of 0.3. Buildings that run all the time, such as tertiary-care hospitals, will tend to have electric LFs of 0.5–0.9. The higher the LF, the less energy cost influence factors such as demand or peak demand have on total energy cost.

Demand or peak demand influences on fuel costs can often be reduced by:

- Rescheduling Loads. If some equipment can be rescheduled to run at night, instead of during the day when all lights are ON and heating or cooling is ON, demand can be reduced.
- Installing More Efficient Equipment. If more efficient equipment is installed, often the total power is reduced, and demand impacts of that equipment can also be reduced.
- Downsizing Equipment. If some energy equipment is larger than needed, such as air-conditioning equipment, demand can be increased more than necessary. If equipment is being replaced and a smaller size (lower installed power) will work, demand effects can also be reduced.
- Thermal Storage. Although thermal storage systems are not discussed directly in this book, thermal storage can be used to make thermal flows for heating or cooling more level, and thus reduce demand.
- Dispatchable Load Shedding. Although this is a fairly complicated item to consider, due to contractual factors, some utilities offer special rates where, if you can shut off some large systems that use a lot of power during a few key hours each year, lower electric prices can be obtained.

Improve Power Factor

Air-conditioning power is delivered as a 60-cycle-per-second sine wave. Magnetic loads such as motors and transformers cause the voltage and current sine waves to be out of phase. This can create a number of problems, including reducing the power capacity that the building electrical distribution system can carry and voltage drops in the building. Lower voltage reduces motor operating efficiency. Many utilities measure reactive power and poor power factor can result in additional charges on the utility bill.

Generally, a good power factor would be greater than 0.9. Power factor correction is usually considered by the building owner to eliminate utility penalties or to increase the load capacity of the electric service, which is diminished by poor power factor.

Power factor can be corrected by installing corrective capacitors, either at the source of the problem or at the building service entrance. Local capacitors benefit building distribution but require the owner to determine the sources of the poor power factor. "Intelligent" capacitor systems at the service entrance can stage automatically to respond to conditions in the building. These systems should be specified by a qualified electrical engineer to ensure that there are no problems with solid-state devices present in the building.

Load Factor

Load factor (LF) is a ratio of available energy to what is actually used. It is a useful tool to gage the consumption characteristics of metered energy over a period of time.

The most easily visualized example of LF is a hospital. There, most loads (the emergency room, hallways, HVAC in patient rooms—occupied or not) are ON at all times; the only exceptions are small areas like administrative offices. Hospital loads are fairly steady and relatively flat. As such, their load factors tend to run from the high 80% to low 90% range. Other building types can have great swings in their energy draw. Think of the office building that is unoccupied all weekend and then has all the lights and HVAC equipment started up at once to get the building running at 7:00 a.m. on a Monday morning before settling into normal operating mode. Such a building would have very low usage followed by a step peak and then a moderate usage level in a repeating pattern throughout the month. The LF in such a site is considerably lower.

Load factor is calculated with the following formula:

$$\text{LF (percent)} = \text{Total kWh for Period} / (\text{Peak kW Demand} \times \text{\#Days in Period} \times 24 \text{ h/day}) \qquad (2)$$

For example:

$$\text{LF} = 41{,}600 \text{ kWh} / (105 \text{ kW} \times 30 \text{ days} \times 24 \text{ h/day})$$
$$= 41{,}600 \text{ kWh} / 75{,}600 \text{ kWh} = 55\%$$

This load factor indicates that the energy consumption for the billing period used by the building is 55% of the total available energy (at the 105 kW peak level).

(Note that LF can also be computed for other types of energy such as district steam, or anything else that where utility or provider is charging on a demand-based rate.)

Energy that is billed on a demand rate structure rewards facilities for improving or increasing their load factor, or conversely one pays more when their load factor is lower than it need be.

Just as management should track consumption over time, LF should also be examined. One should first compare the site's LF to that of similar buildings in their same geographic or climatological area (since LF varies between building types and based upon weather loads). If a building's LF is significantly lower than what is expected an investigation should be performed to determine why. The solution may be as simple as resetting start-up and operating schedules, or it could be a problem with some major energy-using piece of equipment. Additionally, LF should be continued to be computed and looked at on an ongoing basis each month in order to spot problems that arise going forward. (Note, however, that there will be slight swings in LF within a building during different times of the year, due to varying weather and usage loads.)

The level of value to looking at this LF parameter increases in rates with higher demand charge components.

OPERATIONS

Reduce Operating Hours

Equipment uses the least amount of energy when it is OFF. Only operating equipment when needed should be the first area assessed for potential savings. For effective equipment scheduling, the needs of the occupants must be quantified first. These include normal occupied hours, special events, needs during reduced occupancy, etc. Energy use is minimized by tailoring the system operation to match the building needs.

Reduce Equipment Operating Hours

Revising operating schedules to reduce the operation of equipment is one of the most cost-effective ways of reducing energy use in a building. Scheduling can often be applied to the following systems:

Lighting
- Interior lighting scheduled to match occupied hours and possible security needs
- Exterior lighting scheduled for occupancy hours with reduced lighting for security during unoccupied hours

Ventilation Fans
• Schedule to match occupied hours

Outdoor Air Introduction
• Schedule to match occupied hours and avoid outdoor air introduction
during morning warm-up or occupied fan operation to maintain unoccupied setpoints. Alternately, use outdoor air during unoccupied periods for free cooling and/or precooling.

Chilled-Water Systems (Plants and Distribution)
• Outdoor temperature enable
• Time of year scheduled

Hot-Water Systems (Plants and Distribution)
• Outdoor temperature enable
• Time of year scheduled

Service Hot-Water Systems
• Schedule recirculation pumps to occupancy schedule

Coordinating the equipment operation with the building use schedule
requires some planning and coordination with building use and either
manual enabling of systems or provisions to schedule the system
through a control system.

Consideration for how the building is conditioned and lighted during postoccupancy might need to consider accommodations for
reduced occupancy for activities such as cleaning, special events,
occasional intermittent occupancy, etc. Space temperatures and outdoor air ventilation levels might not need to be maintained to occupied levels during cleaning periods. In some instances, lighting might
be used by cleaning staff to track progress through a building by starting with all lights ON and shutting OFF lights as areas are cleaned.
Unoccupied building condition requirements, such as temperature and
humidity, might impact the ability to limit off-hours operation. For
example, it might be desirable to maintain the building pressurization
to avoid humidity intake in some instances. Understanding the needs
and uses of the systems for different purposes is essential in minimiz-
ing the operation of energy-using systems.

Savings Calculations. The savings calculations vary with the type of system being scheduled. The basis for the savings calculations are the change in hours of use multiplied by the energy use per hour of the system. Constantly loaded systems like lighting or constant-volume fans use energy at a fixed rate giving the simplest calculations.

Annual Energy Savings = Energy Use per Hour × Hours per Year Now OFF

Systems with varying loads require an estimate of the total load or an estimate of average load over the period that will be scheduled OFF. Loads might be estimated based on day and night differences, on heating and cooling differences, or need to be made seasonally with weather.

Annual Energy Savings = Energy Use Per Hour at Load Level 1
× Hours Per Year Now OFF as Load Level 1
+ Energy Use per Hour at Load Level 2
× Hours per Year Now OFF as Load Level 2
+ Energy Use per Hour at Load Level 3
× Hours per Year Now OFF as Load Level 3
+... as needed to define all load levels

It should also be noted that reduction in internal heat gain (i.e., by decreasing lighting power consumption) will increase the building heating load and decrease the cooling load.

Cost Considerations. Manual control of system scheduling can be low-cost and effective when the system operation is incorporated into regular workflow. Existing controls can often be augmented with additional schedules for the cost of labor. Automated control can be added for scheduling with hardware and labor costs.

Adjust Space Temperature and Humidity Setpoints

ASHRAE Standard 55-2010 defines the range of temperature and humidity at which most people are comfortable (ASHRAE 2010a). Obviously, lowering the thermostat in the heating season or raising it in the cooling season saves energy. A good rule-of-thumb is 1% per °F for heating savings and a greater percentages for cooling savings. However, setting the building outside of occupant comfort range will negatively impact their performance, which is not the objective in building energy efficiency. Overheating or overcooling the building is wasteful. Simultaneous heating and cooling is another potential cause of wasted energy.

Once the building is set at proper temperatures during the occupied period, make sure the temperature is set back in the heating season and up in the cooling season when the building is unoccupied. There are 168 hours in a week and most buildings are only occupied 40–50 hours per week. There are great savings opportunities in reducing the operation of building systems when no one is in the building. For most building systems, the setback should be 10°F–15°F. Use less setback for heat pump systems as they have less ability to recover than other systems. If the building has an EMS system, make sure it is using optimal start/stop sequences for the air-handling units.

Humidity savings are more difficult to calculate. Energy savings can sometimes be achieved by raising chilled-water temperature in large buildings and using less reheat, provided this action is coordinated with the settings on the terminal units. However, this is not always a good way to save energy. The higher the chilled-water temperature, the less dehumidification occurs, and the potential result is occupant discomfort. The goal of this handbook is energy efficiency without sacrificing occupant comfort.

For buildings with electric humidification, ultrasonic humidification should be considered.

Hot-Water and Chilled-Water Reset

Hot-water and chilled-water reset should be implemented and adjusted to maximize energy savings. Hot-water reset reduces short cycling of the boiler system. Chilled-water reset can improve chiller efficiency and reduce reheat. As noted above, the chilled-water temperature must be cool enough to provide proper dehumidification.

BUILDING ENVELOPE

Heat gains and losses through the building envelope create part of the need for heating and cooling of buildings. The net heat loss from a building during the heating season is the space heating load. Space heating requirements come mainly from heat lost through the building envelope and from the need to heat any outdoor air that comes into the building. Sunlight and internal heat gains from equipment help reduce the heating load.

Similarly, the net heat gain of a building in the cooling season is the space cooling load. Space cooling requirements come mainly from internal heat gains and net heat gain across the building envelope and from sunlight through windows and glass. There is a humidity load also that results from any humid outdoor air that enters the building, people in the building, and possibly from other water sources in the building.

The building envelope can have major impacts on the amount of heating and cooling a building requires, so improving the envelope can be important in some cases. For most existing commercial buildings there are only limited opportunities to improve the envelope. Reduction of heat loss in northern climates and reducing the balance point of the building can often lead to cooling energy penalties, due to higher internal heat gains in commercial buildings than in residential buildings. The balance point of a given building is the outdoor air temperature at which the internal heat gains of the building equal its heating load. Reductions in major air leakage paths are important, as excessive outdoor air leakage influences both heating and cooling energy requirements. Reductions in heat gains are important in areas of the country where commercial buildings have air-conditioning. Reductions in heat losses are important in colder regions.

Reduce Heat Conduction Through Ceilings and Roofs

Energy flows affected by insulation levels in ceilings and roofs in commercial buildings are complex. Heat given off by lighting and other equipment in commercial buildings tends to reduce heating needs and increase cooling needs. Reducing heat conduction through ceilings and roofs, usually by adding insulation, reduces heat flows into and out of the building. This reduced heat flow reduces heating and usually causes an increase in cooling energy use, so a balance must be sought for buildings that are cooled and heated.

Buildings with pitched roofs can have insulation added to the ceilings. Attic spaces under pitched roofs are required to have ventilation, so moisture issues should not be an issue. Insulating attics is relatively low-cost. For low-slope roofs, at the time a building must have the roof replaced, insulation can be added above the roof at a reasonably low incremental cost. If existing insulation has water damage, all the insulation may need to be replaced.

In general, adding insulation below a low-slope roof has moisture risks. Insulation can be added below a low-slope roof even when the roof does not need to be replaced, but potential moisture issues must be evaluated carefully. If there are any significant moisture sources in the building, moisture issues are likely to occur, and unless there is either a tight seal between insulation and the roof deck, or some air leakage (or venting) above the insulation, moisture problems may still occur.

Readers may wish to consult the residential insulation fact sheet (found at www.ornl.gov) (EERE 2008) for more information on insulation, although the recommended insulation levels should be approached with caution for commercial buildings.

The best way to calculate potential energy savings for ceiling or roof insulation is to use an energy simulation program, and an energy professional should be consulted to obtain such results. A case study on one building that increased roof insulation from R-3 to R-17 can be found at www.blackhillsenergy.com (BH 2010), where for a total investment of about $122,000, the company achieved a savings of $11,000/yr in natural gas heating costs in a northern climate.

Reduce Solar Heat Gain through Roofs

Solar heat gain through roofs can have a large impact on space cooling loads. The amount of solar energy absorbed depends on the insulation, roof solar reflectance, and roof area. For an existing building, the solar load can be reduced by increasing the solar reflectance of the roof. The reflectance of a roof can be reduced by applying a special reflective coating, but decisions about roofs are usually made when the roof must be repaired or reroofed. At the time of reroofing, the top surface can be selected to be reflective.

The ENERGY STAR® program has requirements for commercial building roof products, with a minimum solar reflectance requirement (e.g., greater than 0.65 out of a range from 0–1), among others (EPA 2008a). When considering a roof with increased solar reflectance, the ENERGY STAR partners can be checked as a starting point. ASHRAE currently does not promote reflective roofs outside climates zones 1–3 (see map in Appendix A) for new construction (based on roof insulation allowances in Standard 90.1 [ASHRAE 2010c]), but the calculation method below can be used to verify whether reflective roofs might still be of benefit in climate zones 4–7. Buildings with high internal loads

(Type III in ORNL-6527 below [Griggs et al. 1989]) can often benefit from a reflective roof, even in cooler climates.

Green (vegetative) roofs are becoming popular in urban settings. They reduce the "heat island" effect and can provide usable space for the building occupants. However, vegetative roofs are not the most cost-effective means to reduce solar heat gain and should only be considered if there are other reasons for wanting a vegetative roof. They also have structural and hydraulic impacts on the building.

Savings Calculations. The savings calculations for reducing solar heat gain through roofs are complicated. The ENERGY STAR program coverage of reflective roof products (EPA 2011b) has a link to the Roof Savings Web site Calculator (RSC) (ORNL/LBNL 2011). The RSC, Beta Release v0.9, is an industry-consensus, Web-based roof savings calculator for some commercial and residential buildings, using whole-building energy simulations (www.roofcalc.com) (ORNL/LBNL 2011). The AtticSim tool is for advanced modeling of modern attic and cool roofing technologies. (No cost for use.) One tool that is generally used is the RSC for savings calculations for this measure, although the only commercial building types covered at this time are offices, warehouses, and big-box stores.

For those who wish to understand more about the savings calculations, an Oak Ridge National Laboratory report, *Guide for Estimating Differences in Building Heating and Cooling Energy Due to Changes in Solar Reflectance of a Low-Sloped Roof* (Griggs et al. 1989), can be consulted. Also, background papers on the RSC Web site provide more theoretical detail.

If calculations are needed for a building that the RSC does not handle, the methodology in ORNL-6527 (Griggs et al. 1989) should be used.

Installation Costs. Installation costs for reflective roofs should be obtained from potential installers, as the costs can range widely ($0.75/ ft^2 of roof and up for coatings; up to $3/$ft^2$ and more for new reflective membrane).

Reduce Heat Conduction through Walls

Reducing heat conduction through walls is generally not cost-effective. This measure should only be considered when finishing interior walls. For example, if studs and gypsum wallboard are being added to exterior walls, it will probably be cost-effective to install insulation.

Control Solar Heat Gain through Glazing Areas

Most facilities have varying exposures that contribute to imbalances in heating and cooling. Solar gains can be especially troublesome for cooling as shown in Figure 5-2. Reflective window film technology has advanced significantly and can be applied to create greater shading coefficients to reduce unwanted solar loads.

Window films can benefit facilities in a number of ways:

- Lower energy consumption: Films can decrease both cooling and heating loads by limiting the amount of solar gain due to visible and infrared light (the solar heat gain coefficient—SHGC) and reducing heat loss (U-factor), respectively
- Increase comfort: Related to previous goal, human comfort can be increased by eliminating cold or hot spots
- Reduce glare: Another comfort parameter, films can reflect visible light, thus limiting internal light levels

The analysis of the energy and economic effects of this technology are fairly complex and should be left to a professional. However, buildings with a high percentage of glass in the building envelope, and those that experience a significant imbalance of (most importantly) cooling

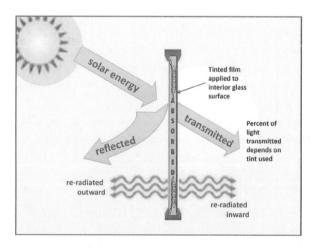

Figure 5-2. Impact of solar energy on glazing.

and (to a lesser extent) heating loads are the best candidates for this type of retrofit. The existing type of glazing is another important factor; buildings with single pane windows and those with older (non low-e) windows will be better candidates. Additionally, the application of window film as a measure should be evaluated, taking into consideration the property's capital plan for the curtain walls.

The key to understanding film performance is the realization that only so much can happen when light hits the window. Upon impact, solar energy can be reflected, absorbed, or transmitted. In addition, the manner in which the three spectra of light interact with the film will also determine performance. To reduce solar heat gain, the entry of visible and infrared light into the interior needs to be minimized. The most straightforward way to accomplish this is to reflect the light outward. The simplest films (i.e., those that are pure metal, like aluminum) reflect all light indiscriminately, which yields good solar performance but results in a shiny exterior look since visible light is reflected as well. For most of the metalized films, the level of solar heat gain reduction is directly proportional to this "shine" factor. Neutral films overcome this by including a dyed layer of polyester which increases the absorption of the film, thus reducing transmitted energy. The balancing act of light gives way to three of the most common aesthetic considerations when selecting a film: the amount of visible light transmitted, the intensity of the external reflectivity, and color. Visible light transmission (VT) seems to be of greatest interest since most film product names include a number which represents this variable.

The classes of high-performance films (namely, low-e, spectrally selective and ceramic) have a more complex array of elements in their coatings and can block much of the infrared and some visible light yet remain relatively clear (no tint) and nonreflective (shiny). As with the neutral product, the energy balance in this case is maintained by increasing the absorption of solar energy. These films are also the most expensive. Based purely on solar performance, this class of films would be the best choice.

The high absorption phenomenon of many of these films gives rise to the necessity of specific product selection based on the installation site. Installers refer to this as film/glass compatibility and it is an issue that requires careful consideration. To prevent the occurrence of glass breakage due to thermal stresses induced in the glass, a detailed analysis of existing flat glass needs to be conducted. Information such as the thickness, surface area, type, orientation, number of panes, any preexisting light modifications (i.e., glass that already includes some type of coating) and any imperfections (that can act as stress concentrations) should be collected before product selection. Another failure mode for multiple paned assemblies is the breakdown of the internal glass seal. In this failure, the seal is stressed by the different rates of expansion of subsequent layers of glass. It has also been documented that the addition of an after-market film on glass that already contains some type of light-altering coating can actually decrease the overall performance of the assembly.

User-operable blinds, shades, or curtains on all windows will give occupant shielding from potentially hot (or cold) window surfaces and thereby minimize radiative energy gain (or loss) by occupant. Shades reduce the need for air-conditioning by blocking and reflecting solar heat, resulting in a 10%–30% reduction in heat gain according to simulation by T.C. Chan Center for Building Simulation and Energy Studies at University of Pennsylvania (2008). Best practice would be to use active interior or exterior solar heat gain control devices. These do not require human intervention but are expensive to install.

Window replacement with low-e insulating fenestration and thermal break frames will save energy but is not cost-effective. If windows are being replaced as part of a building remodel, the incremental cost of installing more efficient windows should be evaluated.

Reduce Infiltration

Infiltration is the loss of energy by movement of conditioned (heated or cooled) air through gaps between building envelope materials or holes/cracks in those materials. Typically, this occurs around window and door frames, as well as where services such as piping and electrical or communications enter a building. This is different from ventilation, in that air taken in or exhausted as part of the ventilation is done so by design or intent, and is (or should be) a controlled process. Infiltration, by contrast, is both uncontrolled and unintended. As such, this is considered an energy loss that should be addressed. It also often causes an imbalance of the temperatures inside the building, which in turn most often results not just in discomfort for those in the local area where the infiltration is occurring, but for many other parts of the building, as the operators in many cases adjust the HVAC distribution controls in an attempt to overcome this. That in turn overheats or overcools other areas of the site.

Infiltration can account for up to 40% of the heating and cooling costs in older buildings. Newer buildings should have been constructed with continuous air barriers.

The most common (and simplest) solution to infiltration, once the source has been identified, is caulking and weather-stripping. Caulking is used to fill or close gaps between two stationary surfaces, i.e., around window and doorframes, between the pane of glass and the window frame, and at penetrations into the building such as piping. Weather-stripping is used wherever two moving surfaces meet, i.e. on a double-hung window—at the top of the window, at the bottom, and in between the two panes, as well as to fill the gaps around doors. On doors, one needs to focus not just on front entrance doors, but on all service doors and roof doors as well. In multistory buildings, infiltration at the lower portion of the building will cause what is referred to as *stack effect*. This is unconditioned air entering at the bottom and pushing the conditioned air up the building (like in a stack or chimney), forcing the conditioned air out all the other holes and exacerbating the drafts that make occupants feel uncomfortable.

When caulking, it is advisable to use a silicon-acrylic mixture type of caulking. While latex caulks will do the same job, they only last about 5 years, and given that the majority of the cost of this measure is labor, it makes sense to use a silicon-acrylic mixture type which will last for 20+ years. There are various types of weatherstripping; which is chosen depends largely on the size of the gap to be sealed and level of activity for the opening. Common effective types include V-seal, tubular vinyl, and reinforced tubular vinyl. It is advised to stay away from foam and felt type as these are porous (letting in the air we are trying to keep out), and deteriorate more quickly. It should be noted that door sweeps (a specialized version of weatherstripping) should be used to seal the larger gaps that exist at the bottom of doors.

Improperly operating door checks should also be repaired or adjusted to reduce the volume of air coming in or escaping. Adding vestibules is a more costly, but effective way to reduce infiltration losses.

Lastly, new windows can also be utilized as a method to reduce infiltration, as these will most often be tighter than those they are replacing. It should be noted, however, that the return on investment on new windows from energy savings is generally very low.

Savings Calculations. The savings calculation for reducing infiltration is:

$$\text{Annual Savings} = ([\text{Average cfm/ft Crack} \times \text{Total Crack Length} \\ \times 1.08 \text{ Btu/}\{h - {}^\circ F - \text{cfm}\} \times 24 \times \text{DD}]/[\text{Btu per Fuel Unit} \\ \times \text{Season Efficiency of Heat Generation}]) \times \text{Current Cost of Fuel Unit} \quad (3)$$

A similar methodology applies for cooling load, but obtaining some of the data is difficult and beyond the scope of this book.

HVAC SYSTEMS

Reduce Excess Ventilation

ANSI/ASHRAE Standard 62.1-2010 prescribes the amount of ventilation air required for a building (2010b). Generally, 15–20 cfm per person is required. Since the ventilation system in a building may have been designed for maximum occupancy, there is a good chance too much ventilation air is being introduced most of the time. For buildings where the occupancy is consistent over the course of the day, the amount of outdoor air can be measured and adjusted accordingly. Spaces such as auditoriums can be fitted with two position controls that bring in a small amount of outdoor air most of the time, and a greater amount of outdoor air during an event. For spaces with highly varied occupancy, CO_2 controls on ventilation air should be considered.

When changing from constant-volume to variable-air-volume systems, care must be taken to ensure that the proper amount of outdoor air is being brought in as the total air delivered to spaces is modulated.

Economizer cycles are now required in most jurisdictions. This has created a potential problem, especially for smaller buildings. Air handlers and packaged rooftop units formerly had two position dampers. There was one position for occupied cycle and the damper was closed for unoccupied cycle. Today, economizer cycles require the use of modulating dampers. These dampers are supposed to close when the building is unoccupied. Often, they do not close because of poor damper construction, faulty linkages or actuators, or because faulty programming causes the damper to return to minimum position instead of closed in unoccupied mode. When properly implemented, combining economizers with demand control ventilation and a setback thermostat can result in significant energy savings in most climate zones.

Improve Chiller Efficiency

Chiller tower system design often optimizes for maximum load with the cooling towers and the condenser-water pumps sized and set up for full-load operation and design conditions. While the chiller efficiency may be good at full load, the chiller is usually more efficient at part loads, and the rest of the components may not be set up to operate the overall plant as efficiently as possible at less than full load. The total cooling plant will operate at part load most of the year. Chiller efficiency is maximized by having the highest possible chilled-water temperature and the lowest possible tower water temperature, within the operating capabilities of the chiller. Analyzing potential energy savings from chiller/tower system modifications usually requires special expertise to assure all influencing factors are considered, so an experienced chiller system professional should usually be consulted.

One important need is to make sure the chilled-water and condenser-water sides of the chiller are not experiencing fouling from bad water conditions. If the heat transfer surfaces in the condenser or chiller tubes are experiencing excessive fouling or scale buildup, efficiency can be reduced dramatically. Chillers can fail from excessive fouling, requiring expensive replacement. A good water treatment program is a necessity for efficiency. Also, if debris, microbiological growth, or scaling are building up in the tower basins or piping legs, special cleaning periods may need to be set aside each year or biocide programs may need to be instituted to eliminate buildups that may cause problems.

The combined effective control of chilled-water supply and return temperatures, condenser-water temperatures, condenser- and chilled-water pump systems, cooling tower fans, and use of chilled water in a building can lead to significant cooling savings in many cases. The complex nature of all these systems working together is best served by special control systems to integrate the control of all these components effectively, and chiller control specialists are often needed to diagnose current system operating modes and suggest effective improvements.

Variable-speed pumping can be an effective way to save chilled-water and condenser-water pumping energy, so installation of variable-speed drives on pumps can be an option for energy savings. Some specialists in chiller modifications will convert chillers to variable-speed drives, but there are not many of these specialists available, and converting a chiller to variable speed will often void any warranty remaining. Only specialized firms with expertise in this area should be used, and special warranties on any such conversions should be considered.

Improve Boiler or Furnace Efficiency

There are a myriad of reasons that a building's boiler or furnace efficiency may not be operating at its optimal efficiency level. If below this level, addressing it may be the most significant or cost-effective energy saving opportunity at the site. The first step is to have a combustion efficiency (CE) test done on the unit.

Combustion efficiency is a measure of how well a boiler-burner (or furnace-burner) turns fuel, oil, natural gas, propane, and coal into usable heat. This is determined, usually by an experienced and qualified boiler specialist, by taking a series of measurements through a small (5/8 inch) hole in the stack of the exhaust gases. Measurements include net stack temperature, percent CO_2 (or oxygen—O_2), stack draft, smoke spot (on oil or coal systems), or parts per million (ppm) of CO (on natural gas systems). From these measurements, not only can the CE level be computed, but specific actions needed to raise CE can often be determined.

The optimal level of CE achievable will depend upon a number of factors, such as how the burner and boiler (or furnace) were designed to operate together, the age of the equipment, and its general condition. As a general rule-of-thumb, the following efficiencies should be attainable:

- 85% for 1–5 year old oil system
- 82% for 1–5 year old gas system
- 83% for 6–15 year old oil system
- 80% for 6–15 year old gas system
- 80% for 16+ year old oil system
- 78% for 16+ year old gas system

Most boiler or furnace units in use should be able to achieve at least 80% CE or better. If the system is operating below that level, action should be taken. It is important to keep in mind that the boiler operates over a load range as the outdoor air temperature varies and the boiler must be tuned for a range of part-load efficiencies.

If limited boiler operation is required in the cooling season, a summer boiler should be considered if the existing boilers do not consist of multiple boilers with high turndown ratios.

The types of measures to be taken to improve a system's CE depend on what is wrong, or indicated by the various measurements taken. They are, however, very often very simple and inexpensive. Some examples include:

- Adjusting the air-to-fuel ratio (by a technician making minor adjustments to the burner)
- Cleaning the tubes, or heat exchange surfaces, on the fire side of the boiler/furnace. (Note that 1/16 in. of soot build-up will reduce CE by 4%). This action requires brushing and vacuuming the surfaces, after which the system should be tuned, to avoid further fouling.
- Lowering the draft level
- Increasing the temperature of the fuel oil supplied to the burner (in heavy oil systems)

In the case that the existing system's CE cannot be brought to achieve an 80% CE level or higher, consideration should be made to replace the system with a new, higher-efficiency unit.

As a general guideline, CE tests should be done at minimum once per year on all boiler of furnace units. Those systems burning large amounts of fuel should be tested more often, from twice a year to once per month.

Savings Calculations. The savings calculations for improving boiler and furnace efficiency is quite straightforward.

$$\text{Annual Savings} = ([\text{New CE} - \text{Old CE}] / \text{New CE})$$
$$\times \text{Normalized Historical Annual Energy Use}^*$$
$$\times \text{Current Cost of Fuel Unit} \qquad (4)$$

* Normalized historical annual energy use is computed by taking the prior year's consumption, breaking out the weather- and non-weather-dependent loads, computing the consumption per degree day (weather measurement parameter) for that actual period, then multiplying that use/DD (degree day) figure by the average number of DDs in a typical year for that location (usually calculated over a 15-year range), and then adding that back with the number to the non-weather-dependent consumption.

Example. The 20-year-old boiler-burner system burning #2 fuel oil in a building is in generally good condition (no excessive leaks or maintenance calls), but it is found to be running at 70% CE. There is a high-stack temperature and a smoke spot of two. The service technician (or engineer) discovers there is a little more than 1/8 in. of soot on the heat-exchange tubes. From that, it is determined that it needs cleaning and a burner tune-up. This will increase the CE to 80%.

The building's normalized (weather adjusted) historical annual energy use is 13,500 gallons of #2 oil. They are currently paying $2.50 gallon.

The savings from this measure is:

$$\text{Annual Savings} = ([0.8 - 0.7] / 0.8) \times 13{,}500 \text{ gallons} \times \$2.50 \text{ per gallon}$$

$$= \$4219$$

Assuming that cleaning costs $600 and a burner-tuning costs $450, then the return on investment for this improvement is 402%.

Improve Air-Conditioner or Heat Pump Efficiency

Air-conditioning equipment (packaged direct expansion cooling) energy use depends on the cooling load and the operating conditions of the equipment. As the outdoor temperature increases, the efficiency of the equipment decreases. Warmer outdoor temperatures require the equipment to use more energy to reject heat to the warmer outdoor environment.

The total packaged cooling system energy use depends on the following:

- The cooling circuit equipment efficiency
- The total cooling load
  - Amount of outdoor air
  - Operating hours
  - Economizer free cooling hours
  - Dehumidification operation
    - Operation only to remove humidity
    - Reheating of air to avoid overcooling the space
- The air distribution energy use
  - Fan efficiency
  - Fan motor efficiency
  - Filter efficiency and cleanliness
  - System pressure losses (duct layout and design)

Improve Existing Equipment Efficiency

The condition of existing equipment can compromise its performance. Improper refrigerant charge leads to operating conditions that reduce energy efficiency. A refrigerant leak reduction/testing program is beneficial in keeping equipment operating at peak efficiency. Similarly, conditions that reduce airflow across the coils evaporator and condenser, pressure drops, damaged fins, etc., reduce the equipment efficiency. It is also important to ensure that condensing unit air discharge is not blocked or diverted back to the inlet. Air recycling results in a significant drop in operating efficiency.

Savings Calculations. Laboratory studies have found correlations in refrigerant charge deficiencies and energy efficiency, but the seasonal impacts vary with loading conditions. Generally, the percentage loss in efficiency is on order to the percentage deficiency of charge (Farzad and O'Neal 1988, Choi and Kim 2001).

Cost Considerations. Neglected equipment can use additional energy on an ongoing basis due to undetected failure or degradation resulting in increased energy costs. A maintenance program should be developed in concert with the building energy management plans. Manufacturers provide guidance on maintenance activities and frequency for their equipment.

Replace Equipment with Higher Efficiency Ratings

Since packaged equipment energy use includes both the cooling circuit (compressors) and air distribution equipment (fans), the efficiency ratings combine both components. The energy efficiency ratio (EER) includes the cooling and the fan energy at a specified testing condition. For equipment smaller than five tons, the seasonal energy efficiency ratio (SEER) is used that incorporates performance at set load levels to approximate operation over the year.

Replacing equipment offers the opportunity to review if the new equipment size should be changed. Oversized equipment can run less efficiently, reduce the ability to remove moisture (condensation on the evaporator coil will not drain as completely as coil/compressor cycling increases), and cause greater electric demand (the equipment connected load is larger and draws more current when operating).

Savings Calculations. For smaller equipment it is adequate to simply take the change in SEER or EER and apply the percentage change to the estimated cooling energy use. In larger systems, isolating the compressor circuit efficiency from other components may be required to identify the energy savings potential.

Cost Considerations. Equipment replacement can be difficult to justify based solely on an improvement in efficiency. The incremental cost of selecting an improved efficiency rating over the standard level can often be justified at the time of replacing older equipment. The best time to replace equipment is often at the end of its service life, at change in tenants or when maintenance costs become excessive.

Use Heat Reclaim Systems

In the heating season, the air being exhausted from the building is warmer than the ventilation air being brought into the building. In the cooling season, the opposite is true and the exhaust air may be less humid as well as cooler. Energy can be saved through heat reclamation systems. There are four types of heat reclamation systems that are generally used: heat wheels, heat pipes, cross flow heat exchangers, and runaround loops.

The heat wheel is the most efficient system and can provide about 80% heat recovery. It can also recover latent heat. The heat wheel requires the intake and exhaust ducts to be run parallel to one another and there is some cross-flow contamination between the intake and exhaust air streams. The wheel rotates slowly between the two ducts transferring energy.

Heat pipes are run between the intake and exhaust ducts and use a contained refrigerant to transfer heat. There is no contamination between the air streams and efficiencies are 70%–80%. Latent heat is not transferred.

Cross-flow heat exchangers use finned metal plates to separate the air streams. Air can be ducted to the heat exchanger and the amount of air passing through it varied with the season. Again, typical efficiencies are 70%–80%.

Runaround loops use coils in the exhaust and intake air streams connected by piping and a pump. Water or a glycol mixture is used for heat recovery. These systems are only 60%–70% efficient but can be used to recover heat from ducts that are not in close proximity to one another.

Savings can be calculated by using the reduced airflow calculations found elsewhere in this chapter and applying the manufacturer's efficiency rating for the particular heat recovery system being evaluated.

HVAC DISTRIBUTION SYSTEMS

Air Distribution Systems

Measures for HVAC air distribution systems have both simple and complex aspects. On the simple side, air-ducting systems should be checked to make sure there are no major holes and that all ducting is connected (no loose connections or disconnected ducts). In addition, any ducts running through unconditioned space, such as ventilated attics, should be insulated. Ducts providing cooling need to be insulated in all cases to help prevent water condensation. Energy is also saved by ensuring that the system is operating, and has been balanced, using the lowest possible system pressure. High-loss fittings in the system should be replaced.

On the complex side, air distribution can become intricately involved in heating or cooling zoning issues and thermostat placement issues that cause heating or cooling systems to run more than actually needed.

Cautions. Changing HVAC airflows can raise contentious issues related to breathing air (indoor air quality, IAQ). Two opposing issues are:

• Concern over IAQ has caused many HVAC systems, especially variable-air-volume (VAV) systems, to be designed so that they do not function correctly and have excessive airflow much of the year. Evaluation of whether minimum flow settings are too high or total airflows are too high (especially during unoccupied hours) is important for determining if airflows can be reduced without impacting IAQ, thus leading to important savings.

• Concern over IAQ has caused some locations to require preoccupancy air purging using outdoor air, so take care not to circumvent local requirements. In addition, airflows should not be reduced to a point where local IAQ requirements are not met.

The main potential energy measure for air distribution involves reducing airflow rates to match actual heating and cooling loads. The primary situation where this would occur is if the HVAC system has constant-volume airflow and can be converted to a variable-flow system.

The other primary means of saving energy involves turning off the air distribution systems when not needed, but that measure is discussed under *Operations: Reduce Operating Hours* later in this chapter.

The savings calculations here are only for air distribution changes and do not include savings from reducing operating hours. If operating hours can be reduced, those calculations should be made under the methods in that section.

To calculate potential energy and cost savings for these air distribution measures, the total annual heating and cooling energy and costs must be added to arrive at total HVAC fuel and/or electricity and costs.

If flow reductions or leak fixes only apply to a portion of the HVAC systems, an estimate must be made of the HVAC fuel and/or electricity and costs for that portion of the HVAC systems.

Savings Calculations:

$$\text{Annual HVAC Electricity} = \text{Annual Heating kWh} \\ + \text{Annual Cooling kWh} \, (+ \text{Annual Fan kWh}) \qquad (5)$$

where annual fan kWh are added if tabulated separately. Annual HVAC Fuel usually equals annual heating fuel kBtu/yr, but also add cooling fuel kBtu/yr, if used.

$$\text{Annual HVAC costs} = \text{Annual Heating Costs} \\ + \text{Annual Cooling Costs} \, (+ \text{Annual Fan Costs}) \qquad (6)$$

where annual fan costs are added if tabulated separately.

Savings factors from Table 5-1 must be selected appropriately, matching the correct table entry to the correct quantity in order to calculate potential savings. These savings factors are reasonably conservative, although some types of conversions to VAV can cause heating energy to increase. Even in those cases, the overall savings estimated

Table 5-1. HVAC Air Distribution Systems Savings Factors

Building and End-Use Type	Fix Leakage or Ducts, Minor Fix	Correct Major Leakage	Moderate Flow Reduction	Major Flow Reduction, VAV or Other
All-Electric Building				
HVAC Electricity	0.5% (0.005)	2% (0.02)	8% (0.08)	15% (0.15)
HVAC Costs	0.5% (0.005)	2% (0.02)	8% (0.08)	15% (0.15)
Electric/Fuel Building				
HVAC Electricity	0.5% (0.005)	2% (0.02)	10% (0.10)	20% (0.20)
HVAC Fuel	0.5% (0.005)	2% (0.02)	0	0
HVAC Costs	0.5% (0.005)	2% (0.02)	8% (0.08)	15% (0.15)

here should be close enough to actual to proceed with final measure prioritization.

$$\text{Annual HVAC Electricity Savings} = \text{Annual HVAC Electricity} \times \text{Savings Factor}$$

$$\text{Annual HVAC Fuel Savings} = \text{Annual HVAC Fuel} \times \text{Savings Factor}$$

$$\text{Annual HVAC Cost Savings} = \text{Annual HVAC Costs} \times \text{Savings Factor}$$

Installation Costs. Installation costs for fixing ducts or leakage are typically low, $200–$500 per location fixed.

Installation costs for moderate flow reductions will depend on the amount of floor area covered by the retrofit or renovation. For moderate flow reductions, installation costs can be estimated as costing

$0.75–$1.25/ft^2 of floor area affected. Major flow reductions can be estimated as $1.25–$2.50/ft^2.

Duct Insulation

HVAC air systems usually have supply and return ducts that carry the air from the heating or cooling equipment to the spaces to be conditioned and return air from the space to the equipment. Uninsulated ducts serve as an energy loss in HVAC systems. In general, it is good practice—and often required by current building codes—to insulate ducts when the building is constructed. ASHRAE Standard 90.1-2010, *Energy Standard for Buildings Except Low-Rise Residential Buildings*, states in section 6.4.4.1.2 (ASHRAE 2010c):

> "All supply and return ducts and plenums installed as part of an HVAC air distribution system shall be thermally insulated in accordance with Tables 6.8.2A and 6.8.2B [of ASHRAE Standard 90.1-2010]."

An exception is that ducts are not required to be insulated if ducts or plenums are located in heated spaces, semiheated spaces, or cooled spaces. Also, for duct runouts less than 10 feet in length to air terminals or air outlets, the rated R-value of insulation need not exceed R-3.5. Ducts in conditioned spaces experience minimal conductive losses and gains since they are exposed to indoor air temperatures. However, these ducts may also require some insulation to prevent condensation on duct walls and to ensure that conditioned air is delivered at the desired temperature.

However, in retrofit applications the cost to add insulation to an already-installed system may be uneconomical. That being said, there are certainly times when adding duct insulation can be economically performed. Adding insulation during a space rehabilitation is one such instance.

Tables 5-2 and 5-3 are excerpted from ASHRAE 90.1-2010, Tables 6.8.2A and 6.8.2B (ASHRAE 2010c). Table 5-3 is used to check if ducts are being considered for insulation, and Table 5-2 is for ducts that only heat or cool (ASHRAE 2010c).

Table 5-2. Minimum Duct Insulation R-Value[a], Cooling- and Heating-Only Supply Ducts and Return Ducts

Climate Zone	Exterior	Duct Location						
		Ventilated Attic	Unvented Attic Above Insulated Ceiling	Unvented Attic with Roof Insulation[a]	Unconditioned Space[b]	Indirectly Conditioned Space[c]	Buried	
Heating-Only Ducts								
1, 2	none	none	none	none	none	none	none	
3	R-3.5	none	none	none	none	none	none	
4	R-3.5	none	none	none	none	none	none	
5	R-6	R-3.5	none	none	none	none	R-3.5	
6	R-6	R-6	R-3.5	none	none	none	R-3.5	
7	R-8	R-6	R-6	none	R-3.5	none	R-3.5	
8	R-8	R-8	R-6	none	R-6	none	R-6	
Cooling-Only Ducts								
1	R-6	R-6	R-8	R-3.5	R-3.5	none	R-3.5	

Table 5-2. Minimum Duct Insulation R-Value^a, Cooling- and Heating-Only Supply Ducts and Return Ducts (continued)

Climate Zone	Duct Location						
	Exterior	Ventilated Attic	Unvented Attic Above Insulated Ceiling	Unvented Attic with Roof Insulation^a	Unconditioned Space^b	Indirectly Conditioned Space^c	Buried
2	R-6	R-6	R-6	R-3.5	R-3.5	none	R-3.5
3	R-6	R-6	R-6	R-3.5	R-1.9	none	none
4	R-3.5	R-3.5	R-6	R-1.9	R-1.9	none	none
5, 6	R-3.5	R-1.9	R-3.5	R-1.9	R-1.9	none	none
7, 8	R-1.9	R-1.9	R-1.9	R-1.9	R-1.9	none	none
Return Ducts							
1 to 8	R-3.5	R-3.5	R-3.5	none	none	none	none

Source: ASHRAE Standard 90.1-2010 (ASHRAE 2010c)
Notes: a. Insulation R-values, measured in (h·ft^2·°F)/Btu, are for the insulation as installed and do not include film resistance. The required minimum thicknesses do not consider water vapor transmission and possible surface condensation. Where exterior *walls* are used as *plenum walls*, *wall* insulation shall be as required by the most restrictive condition of Section 6.4.4.2 or Section 5 of ASHRAE Standard 90.1-2010 (ASHRAE 2010c). Insulation resistance measured on a horizontal plane in accordance with ASTM C518 (ASTM 2010) at a *mean temperature* of 75°F at the installed thickness.
b. Includes crawlspaces, both ventilated and nonventilated.
c. Includes return air *plenums* with or without exposed *roofs* above.

Table 5-3. Minimum Duct Insulation R-Value,[a] Combined Heating and Cooling Supply Ducts and Return Ducts

Climate Zone	Duct Location						
	Exterior	Ventilated Attic	Unvented Attic Above Insulated Ceiling	Unvented Attic with Roof Insulation[a]	Unconditioned Space[b]	Indirectly Conditioned Space[c]	Buried
Supply Ducts							
1	R-6	R-6	R-8	R-3.5	R-3.5	none	R-3.5
2	R-6	R-6	R-6	R-3.5	R-3.5	none	R-3.5
3	R-6	R-6	R-6	R-3.5	R-3.5	none	R-3.5
4	R-6	R-6	R-6	R-3.5	R-3.5	none	R-3.5
5	R-6	R-6	R-6	R-1.9	R-3.5	none	R-3.5
6	R-8	R-6	R-6	R-1.9	R-3.5	none	R-3.5
7	R-8	R-6	R-6	R-1.9	R-3.5	none	R-3.5
8	R-8	R-8	R-8	R-1.9	R-6	none	R-6
Return Ducts							
1 to 8	R-3.5	R-3.5	R-3.5	none	none	none	none

Source: ASHRAE Standard 90.1-2010 (ASHRAE 2010c)

Notes: a. Insulation R-values, measured in (h·ft²·°F)/Btu, are for the insulation as installed and do not include film resistance. The required minimum thicknesses do not consider water vapor transmission and possible surface condensation. Where exterior *walls* are used as *plenum walls*, *wall* insulation shall be as required by the most restrictive condition of Section 6.4.4.2 or Section of ASHRAE Standard 90.1-2010 (ASHRAE 2010c). Insulation resistance measured on a horizontal plane in accordance with ASTM C518 (ASTM 2010) at a *mean temperature* of 75°F at the installed thickness.
b. Includes crawlspaces, both ventilated and nonventilated.
c. Includes return air *plenums* with or without exposed *roofs* above.

Airflow Reduction through Variable-Air-Volume or Leakage Reduction

The main potential energy measure for air distribution involves reducing airflow rates to match actual heating and cooling loads. The primary situation where this would occur is if the HVAC system has constant-volume airflow and can be converted to a variable-flow system (VAV).

The other primary means of saving energy involves turning off the air distribution systems when not needed, but that measure is discussed under *Operations: Reduce Operating Hours*, in this chapter.

Use of carbon dioxide sensors can allow total airflows and ventilation airflows to be managed in ways that are not possible without the sensors. However, introduction of new sensors may introduce operational or maintenance issues. CO_2 sensors can be an effective means of reducing airflows while also managing ventilation requirements.

The savings calculations here are only for air distribution changes, and do not include savings from reducing operating hours. If operating hours can be reduced, those calculations should be made under the methods in that section.

To calculate potential energy and cost savings for these air distribution measures, the total annual heating and cooling energy and costs must be added to arrive at total HVAC fuel and/or electricity and costs.

If flow reductions or leak fixes only apply to a portion of the HVAC systems, an estimate must be made of the HVAC fuel and/or electricity and costs for that portion of the HVAC systems.

Savings Calculations:

$$\text{Annual HVAC Electricity} = \text{Annual Heating kWh} \\ + \text{Annual Cooling kWh (+ Annual Fan kWh)} \tag{7}$$

where annual fan kWh are added if tabulated separately. Annual HVAC fuel usually equals annual heating fuel kBtu/yr, but also add cooling fuel kBtu/yr, if used.

$$\text{Annual HVAC Costs} = \text{Annual Heating Costs} + \text{Annual Cooling Costs} \\ (+ \text{Annual Fan Costs}) \tag{8}$$

where annual fan costs are added if tabulated separately.

Table 5-4. HVAC Air Distribution Systems Savings Factors

Building and End-Use Type	Fix Leakage or Ducts, Minor Fix	Correct Major Leakage	Moderate Flow Reduction	Major Flow Reduction, VAV or other
All-Electric Building				
HVAC Electricity	0.5% (0.005)	2% (0.02)	8% (0.08)	15% (0.15)
HVAC Costs	0.5% (0.005)	2% (0.02)	8% (0.08)	15% (0.15)
Electric/Fuel Building				
HVAC Electricity	0.5% (0.005)	2% (0.02)	10% (0.10)	20% (0.20)
HVAC Fuel	0.5% (0.005)	2% (0.02)	0	0
HVAC Costs	0.5% (0.005)	2% (0.02)	8% (0.08)	15% (0.15)

Data here developed specifically for this book using DOE-2.1e (Gard Analytics 2003) simulations for offices and hospitals.

Savings factors from Table 5-4 must be selected appropriately, matching the correct table entry to the correct quantity in order to calculate potential savings. These savings factors are reasonably conservative, although some types of conversions to VAV can cause heating energy to increase. Even in those cases, the overall savings estimated here should be close enough to actual to proceed with final measure prioritization.

$$\text{Annual HVAC Electricity Savings} = \text{Annual HVAC Electricity} \times \text{Savings Factor}$$

$$\text{Annual HVAC Fuel Savings} = \text{Annual HVAC Fuel} \times \text{Savings Factor}$$

$$\text{Annual HVAC Cost Savings} = \text{Annual HVAC Costs} \times \text{Savings Factor}$$

Installation Costs. Installation costs for fixing ducts or leakage are typically low, at $200–$500 per location fixed.

Installation costs for moderate flow reductions will depend on the amount of floor area covered by the retrofit or renovation. For moderate flow reductions, installation costs can be estimated as costing $0.75–$1.25/ft^2 of floor area affected. Major flow reductions can be estimated as $1.25–$2.50/ft^2.

Variable-Frequency Drives (VFDs)

Variable-frequency drives are one means of achieving variable air volume in the HVAC air distribution system. The savings methods are the same as given previously for VAV or leakage reduction. Costs would be in the major flow reduction range in the table.

Economizer Cycles

Many building economizer cycles are not working properly or are disconnected. Restoring proper function to economizer cycles is usually a very cost-effective measure.

Air Filters

Most HVAC systems that move air have some type of air filter that removes dust and other particles to help keep the air and the equipment cleaner. Air filters introduce a pressure drop that requires energy to overcome, so the presence of an air filter leads to more energy use. For most buildings, a balance must be sought between cleaning the air and using more energy to clean. A dirty filter will slow down airflow and make the system work harder and use more energy, so changing filters regularly is also important for keeping energy use lower.

ASHRAE uses the term *minimum efficiency reporting value* (MERV) in its test standard for air filters, and filtration capabilities for filters are often specified using MERV ratings. The higher the MERV, the more particles can be removed from the air, but usually energy use increases as MERV of the filters increases. Filters are also offered as "low resistance" to signify they should use less energy, but the key factor is pressure drop across the filter at the flow rate of the HVAC system. So evaluation of air filtration for HVAC systems must balance the desire for increased filtration against the potential for reducing energy use by decreasing filter pressure drop. Refer to ASHRAE Standard 52.2 (ASHRAE 2007a) and *ASHRAE Handbook—HVAC Systems and Equipment,* Chapter 24, Table 3 (ASHRAE 2008b), for additional information.

Steam Systems

Reduce Steam Piping Losses (Insulation). Uninsulated pipes, in general, waste energy. This is true for steam piping, hydronic heating pipes, domestic or service hot-water pipes, or even chiller-water pipes. Many times, these are located in unoccupied space and the heat there is of no value. In other situations, they are running through an occupied space, sometimes overheating it and causing uncomfortable conditions, which may in turn result in energy being wasted by opening windows or running mechanical cooling. See Table 5-2 in the Hot Water section for recommended insulation levels.

Reduce Steam Pressure. Steam systems should operate at the lowest pressure that will permit steam to reach all areas of the system. Excess steam pressure wastes energy.

Repair Steam Leaks. Steam leaks can be the source of significant energy waste. These leaks are most frequently found in steam traps and at fittings, pipe junctions, or in valves. The leaks may be due to a loose connection, leaking gasket, or pinhole. Steam losses vary by the size of the hole and the steam pressure as shown in Table 5-5. Size of the leaks, as well as the extent of steam loss, is difficult to measure; hence, to be on the conservative side, even estimating close to the smallest leak possible provides for economics that demand these be addressed as soon as they are spotted. Sometimes leaks located in valves require that they be replaced.

Table 5-5. Estimate of Steam Losses From Leaks

Hole Size, in.	Steam Pressure	Steam Loss, lb/h			
		15 psi	45 psi	115 psi	165 psi
1/16		2	6	15	22
1/8		8	23	60	86
1/4		31	94	240	344
1/2		132	395	1010	1449
1		509	1526	3900	5596

Derived from: *CIBO Energy Efficiency Handbook,* Copyright Council of Industrial Boiler Owners (CIBO 1997).

Table 5-6. Cost of Losses Due to Steam Leaks

Leak Pressure	Hole Size, in.	Steam Loss, lb/h	Hours/ Year	Usage Reduction (Mlb)	Demand Reduction (Mlb/h)	Annual Savings $
165	1/16	22	8760	188.5	0.022	4718
165	1/8	86	8760	754.1	0.086	18,897
45	1/16	6	8760	52.6	0.006	1324
45	1/8	23	8760	201.5	0.023	5055
15	1/16	2	8760	17.5	0.002	451
15	1/8	8	8760	70.1	0.008	1746

Derived from: *CIBO Energy Efficiency Handbook,* Copyright Council of Industrial Boiler Owners (CIBO 1997).

The cost savings in Table 5-6 are derived at $25/Mlb and $1000 Mlb/hr for demand. These are Con Edison's district steam system commercial rates; which are some of the most expensive energy in the continental U.S (Con Edison 2009).

Establish or Maintain Steam Trap Inspection and Maintenance Program

Radiator Steam Trap Maintenance Program. Many facilities do have an existing radiator steam trap maintenance program, while others do not. In buildings where no such program exists one should be established, and in those that do, it should be continually improved. Non-working or improperly working steam traps can wreak havoc to steam distribution and typically lead to steam balancing problems resulting in occupant discomfort.

Every radiator should have the steam trap thermostatic element checked for need of replacement every three to five years. This can be done by: a) existing building staff (the least expensive approach); b) a part-timer or apprentice to the super/maintenance personnel, as the existing building staff may not have time to conduct this task along with all his other responsibilities; or c) an outside contractor (the most costly, and not generally recommended, as this is not a high-skill task that requires the use of contract labor). Before beginning this undertaking, it might be advisable to have a knowledgeable plumber randomly inspect a random sampling of traps to check for wiring or drawing (erosion) of the seats (bottom surface of the trap outlet).

If there is no significant erosion of the trap, seats replace the radiator trap elements (disks). If there is significant erosion of the trap seats, do one of the following:

- Replace all of the trap elements, and where necessary replace the worn seats (these are often threaded/screwed into the trap bodies and can be replaced).
- Replace all of the radiator trap elements with cage or cartridge units. These are devices that are specially designed trap replacement units that are made to act as "pop-in" replacements for existing elements. The advantage of the cage or cartridge units is that they are fitted with a "new" seat that is integral to the unit and that sits on the existing trap body seat and seals the steam bypass leaks.
- Replace all of the trap bodies and their elements (this option should be considered a last resort, and is unlikely to be necessary at many facilities, particularly ones at which the level of maintenance and upkeep of the system is high).

A radiator steam trap maintenance log should be made as to what action was taken in each radiator. After the initial upgrade, log sheets should continue to be kept as additional trap maintenance is done on an as-called-for basis. When management observes that the number of trap maintenance calls have increased (after a few years) this should act to once again trigger another building-wide program.

Main/Basement Steam Trap Tagging and Maintenance Program. Either a continued program, or a new effort of repair/replacement of non-working steam traps, will increase both occupant comfort and overall operating efficiency. A main steam trap tagging and maintenance program should be instituted as soon as possible after completion of the radiator trap program. The first step is to tag all of the traps on the steam mains. A main steam trap master location log should be used as an easy reference for building staff and contractors in identifying all traps. A service contractor can then be brought in every two to three years to inspect all traps and repair (rebuild) and/or replace each trap as necessary. Note that it is likely that, in a very large percentage of sites, some of the traps will still be found to be in good working condition. Any inspections and maintenance actions should be noted on the main steam trap maintenance form.

Hydronic Systems

Variable-Speed Drives. Pumping systems are a common system in many commercial buildings. According to the U.S. Department of Energy, pumping systems account for almost one-fifth of global electric motor load (DOE et al. 2004). In new buildings, current design practices and most building codes require the use of variable flow in both hydronic (heating and cooling) and ventilation/circulation air systems. However, there are many existing and older buildings that were designed and still have constant flow hydronic systems.

In most commercial building applications, the load on hydronic pumping systems is or can be made variable. However, in a large percentage of those applications there is no or inefficient hydronic flow control. Proper implementation of pump/flow control using variable-speed drives (or variable-frequency drives) offers an efficient and sustainable approach to improving flow control while reducing onsite energy use. There are also nonenergy benefits such as improved controllability, improved system performance, and extended equipment life that can be realized through the proper implementation of variable-speed drive technologies.

As the pump speed is reduced (to reduce or vary the flow), the amount of energy required to circulate the fluid is reduced exponentially. Common approaches to flow control utilize some type of control loop feedback, usually a static pressure sensor, located in the system to provide feedback to the variable-speed drive controller. There are many public-domain analysis tools available to help customers evaluate potential energy savings from implementing variable-speed drives on their systems. In addition, most vendors and manufacturers offer analysis tools (some even on their Web sites) that customers can use to estimate energy savings. A good primer and application guide can be found on the DOE Web site (www1.eere.energy.gov) (DOE et al. 2004).

While the energy savings resulting from a variable-speed drive installation are very site and system specific, typically 25% to 35% energy-use savings can be achieved in a well-designed application.

The use of variable-speed drives also help mitigate the common design approach of oversizing equipment and system capacities. In an effort to provide a safety factor in design calculations and also accommodate unforeseen load increases, hydronic flow systems are often oversized. The use of variable-speed drives will allow the pumping system to operate at the needed flow rate, thereby reducing energy use associated with the provision of excess flow.

Savings calculations. A simplified approach to estimating energy savings in a constant-flow to variable-flow application using variable-speed drives is to compare the energy use of the flow configurations. The energy use of the constant-flow pumping case can be estimated using:

$$kWh_{constant\,flow} = \frac{OpH \times BHP_{full\,load} \times 0.746}{\eta_{motor}} \tag{9}$$

where:

$OpHrs$ = pump operating hours
$BHP_{full\,load}$ = brake horsepower at full-flow pump load
η_{motor} = motor efficiency

The energy use of the pumps with a variable-speed drive can be estimated using:

$$KWh_{variable\,speed\,drive} = \frac{\sum OpHrs_i \times BHP_{full\,load} \times (flow_i)^n \times 0.746}{\eta_{motor} \times \eta_{vsd}} \tag{10}$$

where:

$OpHrs_i$ = pump operating hours at load condition i
$BHP_{full\,load}$ = brake horsepower at full-flow pump load
$flow_i$ = flow at load condition i, expressed as a fraction
n = exponent characterizing flow verses power relationship, ranges from 2.5 to 2.7 for most systems (note: while fan laws contain a exponent value of 3, real building systems do not operate at this condition)
η_{motor} = motor efficiency
η_{vsd} = variable-speed drive efficiency (typically 98% or better)

The annual energy savings can then be determined as the difference between $kWh_{constant\,flow}$ and $kWh_{variable\text{-}speed\,drive}$.

For other flow scenarios, such as two-speed pumping, inlet vane control, bypass control, etc., the baseline energy use can be calculated using a modified approach (using a multiple-flow-condition equation such as presented for the variable-speed drive above).

Water-Side Heat Reclamation. Electrical equipment generates heat that needs to be removed from the space and rejected to the outdoors. Typically this is done using mechanical cooling systems, heat rejection, or economizers (if ambient temperatures allow).

In certain applications, it is possible to reclaim or recover that heat and deliver it to other spaces in the building that need it. Consider, for example, an office building that also contains data center spaces. During the winter the office spaces typically need heat; however, the data center generates significant amounts of heat that is rejected through mechanical cooling systems. Other sources of waste heat that could be reclaimed and reused for space heating include steam condensate and exhaust heat recovery. Building operators should evaluate their buildings and consider if there is heat being rejected to the outdoors that could be recovered and used for space heating or domestic hot-water heating.

Savings Calculations. Calculating the savings of heat reclamation is very application specific. But in general the benefit can be estimated as:

$$\text{Heat Recovered} = \text{Heat Available for Recovery} \times \eta_{hr}$$

$$\text{Energy Saved} = \text{Heat Recovered} / \eta_{htg} \qquad (11)$$

where:

η_{hr} =   efficiency of heat recovery device or system (typically between 50% and 70%), expressed as a decimal

η_{htg} =   efficiency of heating system, expressed as a decimal

WATER HEATING

Reduce Hot-Water Loads

The opportunity here varies greatly by the type of application or building use. Buildings, such as hotels, hospitals, multifamily residences, or even ones with restaurants or health clubs in them, are likely to have high-domestic or service hot-water (DHW or SHW) loads. The other extreme is the typical office building in which there is nothing more than a few restroom sinks used for occasional hand washing.

The typical action to be taken is to replace existing fixtures with low-flow devices. For sinks, install faucet aerators (the faucet needs to have threads on the end to accommodate this; some very old fixtures or newer high style ones do not). This keeps the water forceful and spread out providing a useful flow. For showers, install low-flow showerheads. A low-flow showerhead will cut an older style unit's flow from 5–6 gallons per minute (gpm) to 2–3 gpm. It is very important, however, to utilize a nonaerating-type showerhead. While the aerating types will cut the flow to the lower level, the droplets that it sprays are so small that they cool off before they get to the bather, who in turn adjusts the controls to utilize a larger percentage of heated water, thus negating much, if not all, of the savings. It is also advised that facilities purchase higher-quality low-flow showerheads (ones with multiple spray patterns) as these provide a higher user satisfaction and tend to stay in place over time, securing the savings for the site.

The other, sometimes even more significant, opportunity here is to track down and eliminate leaks. These may be found in either the DHW or SHW distribution system/piping or at the end-use fixtures. To understand the impact of even a small leak, consider a sink faucet dripping at (the slowest they rate such) one drop per second. That adds up to 192 gallons per month or in excess of 2300 gallons per year. A leak where there is a three inch smooth stream coming down from the end of the faucet adds up to 1095 gallons per month, and if let to go on for a year it is in excess of 13,000 gallons per year. Most of the time repairing these requires nothing more than a washer and a few minutes.

Reduce Hot-Water Heating System Losses (Insulation)

In general, uninsulated pipes waste energy. This is true for steam piping, hydronic heating pipes, domestic or service hot-water pipes, or even chiller-water pipes. Looking at the example of hot-water pipes, these waste energy in that either they are located in unoccupied space and the heat there is of no value, or they are running through an occupied space heating up a cooled space, that then needs additional cooling. Even in the situation where uninsulated hot-water pipes are running through a heated space, this added heat is uncontrolled and at times creating an overheated and uncomfortable space.

For any pipe that is usually warm to the touch, it is typically cost-effective to add insulation to the piping.

Table 5-7 shows recommended levels of insulation for various types of piping:

Table 5-7. Pipe Insulation Requirements[a,b,c,d]

Fluid Operating Temperature Range (°F) and Usage	Insulation Conductivity		Nominal Pipe or Tube Size (in.)				
	Conductivity Btu·in./(h·ft²·°F)	Mean Rating Temperature, °F	<1	1 to <1 1/2	1 1/2 to <4	4 to <8	≥8
			Insulation Thickness (in.)				
>350 °F	0.32–0.34	250	4.5	5.0	5.0	5.0	5.0
251°F–350°F	0.29–0.32	200	3.0	4.0	4.5	4.5	4.5
201°F–250°F	0.27–0.30	150	2.5	2.5	2.5	3.0	3.0
141°F–200°F	0.25–0.29	125	1.5	1.5	2.0	2.0	2.0
105°F–140°F	0.22–0.28	100	1.0	1.0	1.5	1.5	1.5

a. For insulation outside the stated conductivity range, the minimum thickness (T) shall be determined as follows: $T = r\{(1 + t/r)^{K/k} - 1\}$ where T = minimum insulation thickness (in.), r = actual outside radius of pipe (in.), t = insulation thickness listed in this table for applicable fluid temperature and pipe size, K = conductivity of alternate material at mean rating temperature indicated for the applicable fluid temperature (Btu·in./h·ft²·°F), and k = the upper value of the conductivity range listed in this table for the applicable fluid temperature.

b. These thicknesses are based on energy *efficiency* considerations only. Additional insulation is sometimes required relative to safety issues/surface temperature.

c. For piping smaller than 1½ in. and located in partitions within *conditioned spaces*, reduction of these thicknesses by 1 in. shall be permitted (before thickness adjustment required in footnote a) but not to thicknesses below 1 in.

d. For direct-buried heating and hot-water *system* piping, reduction of these thicknesses by 1.5 in. shall be permitted (before thickness adjustment required in footnote a) but not to thicknesses below 1 in.

e. The table is based on steel pipe. Nonmetallic pipes scheduled 80 thickness or less shall use the table values. For other nonmetallic pipes having *thermal resistance* greater than that of steel pipe, reduced insulation thicknesses are permitted if documentation is provided showing that the pipe with the proposed insulation has no more heat transfer per foot than a steel pipe of the same size with the insulation thickness shown in the table.

Savings Calculations. The savings calculation for insulating exposed piping is:

$$\text{Annual Savings} = ([U_{old} \cdot A_{old} \cdot T] - [U_{new} \cdot A_{new} \cdot T]$$
$\times$ System Operating Hours per Year) $\times$ Length of Pipe to be Insulated (in linear feet) / Btu per Fuel Unit $\times$ Season Efficiency of Hot-water Generator $\times$ Current Cost of Fuel Unit $\hfill$ (12)

Example: Insulation is needed on 5 feet of 1 ½ in. and 11 feet of 2 in. DHW piping.

Total area to be insulated = 16 linear feet

Btu/linear ft/yr savings (for adding 1 in. insulation to a 1 1/2 in. pipe at 70°F delta-T for 8760 h/yr of operation) = 221,628 Btu/linear ft/yr

Btu/linear ft/yr savings (for adding 1 in. insulation to a 2 in. pipe at 70°F delta-T for 8760 h/yr of operation) = 310,104 Btu/linear ft/yr

New DHW seasonal efficiency of boiler plant = 70%
Cost of fuel: $1.25/therm

Savings = ([5 · 221,628] + [11 · 310,104]) / 100,000 · 70%

Savings = 65 therms
Fuel cost savings: $81
Improvement costs:
   Pipe insulation (materials only): $106
   Total costs: $96
Simple return on investment: 77%

Reduce Hot-Water Heating System Losses (Recirculation)

Circulation systems for heating hot water in older existing buildings were often designed as constant flow. Over the past few decades, many of these systems have been converted to variable flow, often using variable-speed drives. For variably-loaded systems with pumps of 5 hp or larger, adding variable-speed drives and control provides an excellent opportunity to reduce pumping and heating energy use. In addition, in a properly implemented system, comfort will improve and overheating is often eliminated. Variable-speed pumping has excellent applications for most commercial buildings.

Reduce Hot-Water Heating System Losses by Temperature Reset

Boiler efficiency is maximized and circulation losses minimized by controlling hot-water heating systems to the lowest possible water temperature.

Reduce Domestic Hot-Water Circulation Losses

In many facilities, DHW is circulated around the building continuously. In facilities that are not 24 hours per day and 7 days per week operations, this is clearly not needed. Even in buildings that are occupied at all times (such as hotels, multifamily buildings, nursing homes), it is not necessary to run DHW recirculation pumps continuously, circulating hot water through the pipes around the building at 120°F–140°F, as is often done. There are a number of strategies for achieving savings, from time clocks, to learning controllers, to aquastats. The simplest and most inexpensive solution that has also been shown to maintain user hot-water quality satisfaction is installing and setting a reverse-acting aquastat on the return line. This will reduce the amount of energy used to provide hot water to the taps, while retaining the quality of hot-water delivery services. In addition to thermal savings (on the DHW generation), there are some electrical cost savings from the reduced run time on the pumps.

In facilities that are not continuously occupied, a combination of timers to cut out the generation system and pumps during unoccupied times and a reverse acting aquastat on the return line (which acts as the lead control during occupied periods) should be considered.

The situations above assume that existing systems are the best to serve the building and stay in place, with some very inexpensive controls being added to produce very high return on investment improvements. There are instances, however, where changes to the generation equipment and delivery systems/methods should be considered. Such instances are generally characterized by either long distances between the DHW generation source and various points of end use, and/or very low and infrequent volumes of DHW consumption. Classic examples of this are a school with a central boiler plant and a few bathrooms located hundreds of feet from the mechanical room. In such cases, even insulating the piping (if that were even possible in an existing building) provides only minor value as the water sits for long periods of time, eventually losing the heat. It has been shown that in such cases there are significant benefits to using distributed, low-volume, instantaneous DHW generators hooked into the localized piping.

Savings Calculations. The savings calculations for these measures are very complex, and too intricate for this guide. Monitored research studies, however, exist that allow one to estimate the level of savings achievable with the most basic of controls in certain types of operation (and by transposing to others). The NYSERDA-sponsored DHW Recirculation System Control Strategies research study (Goldner 1999) demonstrated that as compared to a pump running 24 hours per day, a reverse-acting aquastat installed a few feet upstream of the pump and set at 110°F, with the deadband set at +/– 5°F, and wired into the domestic hot-water circulation pump in the building, saved 10.8% of facilities DHW energy load.

Example: Add DHW Recirculation Controls.

A building's DHW load = 21,388 therms

Savings = 21,388 × 10.8%
Savings = 2310 therms × $1.25/therm
Fuel-cost savings are = $2887
Improvement costs:
    Add new return line aquastat at $250 each (installed) = $250
Total costs = $250
Simple return on investment = 1155%

Use Energy-Efficient Water-Heating Systems

Dependent upon the type, layout, and utilization of the building there may be an opportunity to make this DHW or SHW more efficient. In some parts of the United States there is a prevalence of combined heating and DHW systems. That is, having one boiler that produces either the steam or hydronic heating water for space conditioning as well as creating the DHW/SHW for the building. These systems were put in largely due to the lower first cost of having one piece of equipment rather than two. However, studies and practice have demonstrated that, because such systems are oversized for the individual loads, they result in significantly lower operating efficiencies than ones sized just to meet the separate needs. Even in the example of a combined boiler in the northeast where the building's DHW load is nearly equal to the space-heating load (say a moderately sized multifamily site), the boiler's seasonal efficiency during the summer, when only DHW is needed, may be as low as 40%. Compare that to an efficiency of 80%–90% (high end with condensing boilers) that might be achievable by a separate DHW-only maker, and the savings become clear.

It should be noted that, while there are significant savings to be had from separating out DHW and space-heating generation, this change generally only makes sense when the existing system is at the end of its useful life and/or needs be replaced for other reasons. The life-cycle savings from such a change are quite positive, but generally not high enough to overcome the first cost of putting in the separate DHW system if another generation mode currently exists, even if it is running at low efficiencies. At the point that the existing system is to be replaced, the two loads should be separated and systems designed to individually each meet. That will provide for not only the energy savings discussed, but also from savings from the lower cost of the smaller replacement (heating only) boiler. It is crucial that both these systems be properly sized to meet the actual loads of the building and not just be replaced with a "what was there before" or some rough rule-of-thumb method.

LIGHTING

Lighting accounts for 21% of all energy consumed in commercial buildings in the United States, as shown in Figure 4-1: End-Use Percentages. Energy is saved and electric demand can be reduced by reducing illumination levels, improving lighting system efficiency, reducing operating hours, and using special lighting approaches such as daylighting and task specific lighting. Reduction of lighting energy can also increase building heating consumption and decrease building cooling consumption, since internal heat gains are reduced. However, using lights to heat a building is not an effective means of heating. If the main cooling equipment is to be replaced for any reason, making lighting efficiency improvements at the same time can reduce the required equipment size.

In many cases, the best efficiency improvements will result from combinations of the efficiency measures described in the remainder of this lighting section.

Evaluate Existing Lighting System

Many existing lighting programs are targeted at installation of more efficient luminaries using a one-for-one replacement to save energy.

This approach presumes that the existing lighting design was performed to current standards and that the occupancy of the individual spaces has not changed. It is preferable to first evaluate the lighting system in terms of current occupant needs and task locations. The results of this evaluation are meant to indicate the best course of action to improve both the lighting system and its energy efficiency. Light quality is as important as lumen efficiency in achieving an efficient lighting system. Improvement options would include:

- A total system replacement (e.g., removing a uniform general lighting grid pattern system to one that is task specific)
- Modifying the existing installation:
  - Removing lamps/ballasts where not needed
  - Relocating/grouping tasks/workers so that lighting can be adjusted and unneeded lamps/luminaries removed
  - Evaluating daylight contribution and making appropriate adjustments to the existing system—lamp/ballast removal, adding controls to lower electric lighting during daylight contribution hours
  - Evaluating/improving current cleaning and maintenance program to insure that lighting levels are maintained at design levels especially if lamps are removed or fixtures disconnected—get the most light possible out of the existing system since the energy is paid for
  - Replacing less efficient lamps/fixtures with new technology sources (e.g., T-12 fluorescent lighting with T-8 or T-5 luminaires, replacing incandescent lighting with compact fluorescent or LED)
  - Greater use of controls to match hours of occupancy

The Illuminating Engineering Society (IES) produces the *Lighting Handbook* that covers lighting applications and recommended lighting levels and approaches (IES 2011b). This handbook is very comprehensive (and can also be found online at www.iesna.org/handbook). For specific building types, IES offers *Recommended Practice* publications; e.g. *Office Lighting* (IES 2004) or *Industrial Lighting* (IES 2006), found at www.ies.org.

Reduce Illumination Levels

Reducing illumination levels can be accomplished through such activities as removing lamps in fixtures where there is too much lighting and increasing or improving use of task lighting, accent lighting, or daylighting. One of the examples in the previous chapter discusses a reduction in electric lighting levels in a warehouse that had daylighting panels in the roof, and the lighting system was changed to install task-specific lighting and to make use of the daylighting already present.

Note that dirty lamps and reflectors reduce ambient lighting levels. Improved cleaning and maintenance is required to maintain required light output from fixtures. When illumination levels are reduced, measurements must be taken to ensure that adequate lighting levels are being maintained.

The specific focus of the rest of this section is on removing unneeded lamps or fixtures. Care must be exercised, though, to assure that any unsafe lighting conditions are corrected.

Lighting levels can be reduced by removing lamps or selected fixtures. Hallways are often illuminated much beyond recommendations. In most instances, two lamps can be removed from four-lamp fluorescent fixtures, or one lamp from a three-lamp fixture. Consider using light colored, reflective surfaces/finishes as a way to decrease lighting demand/ requirements. Incandescent lamps should be removed as much as possible. When removing fluorescent or high-intensity discharge (metal halide, multivapor, sodium, and other) lamps (fluorescent or HID lamps), also remove or disconnect the ballast, if reasonably easy, to achieve additional energy savings. The ballasts will continue to consume energy if left connected and the electric power is ON (they might also make more noise).

Savings Calculations. The savings calculations for lighting are fairly simple if the potential interactive heating and cooling effects are ignored. The heating and cooling effects can be ignored for calculation purposes in most cases, but all-electric buildings in very cold climates do have savings degradation issues, because the *heat of light* decreases the building heating load. Increases in heating energy use are often too small to see in heating bills, and any cooling energy savings will act as a *safety factor* in estimating savings. Heating- and cooling-cost changes will work to cancel each other out over a year in most locations. Cooling benefits for all-electric buildings in hot climates will be an additional benefit.

Lighting efficiency improvements almost always lead to electric demand savings, but average electricity costs can still lead to overestimating cost savings, and some savings reduction factors are needed.

The calculations here require counting the fixtures affected and multiplying by the appropriate wattages, as indicated. This approach can also be used for calculating savings due to reducing lighting operating hours.

Calculating Savings by Counting Lamps or Fixtures. If you know the approximate watts per lamp or watts per fixture that will remain OFF, be removed, or be disconnected, the savings can be calculated based on the number of lamps or fixtures and the number of additional hours per year the lamps will be OFF. Be careful about NOT counting savings from lamps that are not working. If lamps are already OFF because they are not working, there will be no additional savings.

The number of hours the lamps are used before the changes should be known in order to calculate the energy saved. The best way of calculating the hours is to look at h/day and days/week the lamps or fixtures were used (e.g., h/yr = h/week × 52). If the lamps were used 12 h/day for 5 days/week plus an additional 6 h on weekends, the h/week would be 5 × 12 + 6 = 66 h/week. H/yr are calculated as 66 × 52 = 3432. (If one wants to add hour for one more day per year, since one year is 52 weeks + 1 day, that is fine.)

Annual Lighting Electricity Savings = (No. of Fixtures/lamps Now OFF
/ 1000) × Watts per Fixture/Lamp × h/yr Now OFF   (13)

(kWh/yr comes from total Watts / 1000 × h/yr)

Annual Lighting Cost Savings = Annual Lighting Electricity Savings (from equation above) × Annual Lighting Electricity Cost / Annual Lighting Electricity Use × Savings Reduction Factor from table   (14)

($/yr comes from kWh/yr saved times total lighting $/kWh = $/yr divided by kWh/yr)

For dimming ballasts, see Figure 5-3 (fluorescent dimming curve) or refer to figures 27-15 A and B in the 9th Edition of the IES Handbook (IES 2000). Savings from dimming lighting is nearly linear. The more lights dimmed, the more energy saved.

For dimming systems, the energy savings equation becomes:

Annual Lighting Electricity Savings = (No. of Fixtures/lamps Now OFF
/ 1000 × Watts per Fixture/Lamp × h/yr Now OFF
+ (No. of Fixtures/Lamps Dimmed / 1000 × Watts per Fixture/Lamp
× [1 – % Light Level] × h/yr Dimmed)   (15)

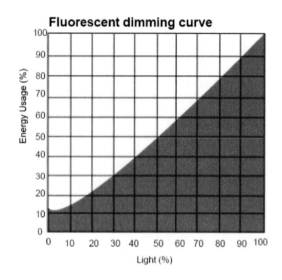

Figure 5-3. Fluorescent dimming curve.
(Photograph courtesy of Lutron Electronics Company, Inc.)

Table 5-8 presents maximum lighting power densities, in terms of watts per square foot, as presented in ASHRAE Standard 90.1-2010 (2010c). These values are the maximum lighting power densities allowed on a building-wide basis. The ASHRAE standard also presents values on a space-by-space method.

Installation Costs. Installation costs for reducing the lamps or fixtures used is typically low, and can be considered zero or equal to the number of hours of work required times the $/h for the workers.

Install Energy-Efficient Lighting Systems

Whether choosing a total lighting system replacement or just replacing parts of the existing lighting with more efficient equipment, care is needed to assure that lighting quality remains acceptable, that efficient lighting control is not compromised, and that expected energy savings meet expectations. As in previous discussions here for lighting, reducing lighting energy tends to increase energy used for heating and reduce energy used for cooling.

Table 5-8. Lighting Power Densities Using the Building Area Method

Building Area Type[a]	LPD (W/ft^2)
Automotive facility	0.82
Convention center	1.08
Courthouse	1.05
Dining: bar lounge/leisure	0.99
Dining: cafeteria/fast food	0.90
Dining: family	0.89
Dormitory	0.61
Exercise center	0.88
Fire station	0.71
Gymnasium	1.00
Health-care clinic	0.87
Hospital	1.21
Hotel	1.00
Library	1.18
Manufacturing facility	1.11
Motel	0.88
Motion picture theater	0.83
Multifamily	0.60
Museum	1.06
Office	0.90
Parking garage	0.25
Penitentiary	0.97
Performing arts theater	1.39
Police station	0.96
Post office	0.87
Religious building	1.05
Retail	1.40
School/university	0.99
Sports arena	0.78
Town hall	0.92
Transportation	0.77
Warehouse	0.66
Workshop	1.20

a. In cases where both a general building area type and a specific building area type are listed, the specific building area type shall apply.

Energy efficiency of lighting is usually evaluated as lumens per Watt, lighting power divided by electric power. More efficient systems can deliver more light for the same energy when compared to less efficient systems. Depending on the tasks in a space, lighting might also be made more effective by delivering more light to the task. One quantity that impacts potential tasks is the color rendering index (CRI) of the light source. The CRI provides an indication of how well colors can be distinguished or perceived with a specific lighting source. The CRI is obtained as the mean value of measurements for a set of special color tests, and has a value of 1–100, where higher means better. Table 5-9, efficacy and CRI for common light sources, provides an indication of typical ranges of lumens/W and CRI.

A CRI < 50 means the light source is not suitable for tasks where ability to distinguish color can be important. These values are typical ranges, but some specialty lamps may provide values outside the ranges shown. Specific products should be checked for actual efficacy or CRI.

Since lighting energy savings is often achieved by turning off lights whenever they are not needed, special care must be taken to assure that lighting systems can be turned OFF as much as possible, in a manner that does not disturb occupants and without reducing the life of the lamps excessively. Many retrofits are occurring where lighting systems that could not be turned ON and OFF at fairly regular intervals without introducing operational or other problems are replaced with systems that can be turned ON and OFF more often. Many HID systems cannot be

Table 5-9. Efficacy and CRI of Common Light Sources

Category	Lumen/W	CRI
Incandescent	10 to 35	+95
Mercury vapor (HID)	20 to 60	20 to 40
Light-emitting diode (LED)	20 to 110	65 to 97
Fluorescent	40 to 100	60 to 90
Metal halide (HID)	50 to 110	65 to 90
High-pressure sodium (HID)	50 to 140	20 to 30
Low-pressure sodium	100 to 180	Low ~5

HID means high-intensity discharge.

turned ON and OFF regularly because they take several minutes to restrike. Bilevel ballasts are available that permit operation at lower lighting levels when the space is unoccupied, but retrofit to a fluorescent lighting system is more cost effective in most instances.

Savings Calculations. The savings calculations are, again, fairly simple, if the potential interactive heating and cooling effects are ignored.

$$\text{Annual Lighting Electricity Savings, kWh/yr}$$
$$= (\text{Watts for Old Lighting} - \text{Watts for New System}) / 1000$$
$$\times \text{ h/yr System is Used} \tag{16}$$

LED Systems

LED systems are a special case in that the technology is changing so fast that savings and costs are hard to pin down. LED systems are currently comparable to fluorescent lighting in efficacy, but the system lifetimes are expected to approach 50,000 h, as compared to 10,000–20,000 h for fluorescent. There are ongoing issues about standards for testing and expected lifetimes, so consideration of an LED system becomes more complicated. Due to the high rate of change, a special Web site has been created to allow interested parties to check what is currently known and what products are available (www.lightingfacts.com) (DOE 2011b).

Lighting Control

Reduce lighting usage through management and controlled systems—consider bringing the lighting control protocols for the building up to Standard 90.1-2010 (Section 9.4.1) (ASHRAE 2010c) standards.

Interior Lighting. The following are options to control and reduce interior lighting usage:

• Reduce operating hours for lighting systems through the use of controls and building management systems. This includes the use of shut-off controls such as time clocks.
• Use reduced lighting levels, including OFF, when spaces are unoccupied, during nighttime hours, for restocking, cleaning, and security. Whenever possible, move restocking and cleaning operations to daytime.
• Use occupancy, vacancy, or motion sensors. Wherever applicable, these sensors should either be manual-ON or turn lighting-ON to no

more than 50% of lighting power. Occupancy sensors automatically turn OFF when occupants leave a space. These sensors can reduce lighting-electricity use from 15% to 60%, depending on the use and size of the space. The latest sensor solutions use radio frequency (RF) technology, allowing them to be installed in minutes with no additional wiring and making them an ideal choice for retrofit applications. Occupancy sensors are best suited for private offices, conference rooms, restrooms, and classroom spaces.

- Dimming controls: Use controls to allow for multilevel or dimming control of the lighting in the appropriate spaces. Dimming lights saves energy. The savings is nearly linear (i.e., dimming fluorescent lights by 30% saves 25% in electricity, dimming them by 50% saves 40% in electricity, and so on). Furthermore, users seldom require maximum light levels and studies show that allowing users to adjust illuminance for different tasks saves 35%–42% lighting energy. Plus, it's highly desirable and makes tasks seem less difficult (Derived from Figures 27-15A and 27-15B in *IES Lighting Handbook*, 10th edition [IES 2011b] and *Individual Lighting Control: Task Performance, Mood and Illuminance* [Boyce et al. 1999]). Using strategies like these in combination, it is typical for buildings to cut their lighting energy usage by 60% or more. Additionally, because lights emit heat, lighting control can reduce HVAC demand. As a rule-of-thumb, for every 3-watt reduction in lighting power, there is a 1-watt reduction in cooling load.

- Digitally addressable dimming ballasts: Digitally addressable dimming ballasts are the building blocks of lighting systems that are fully controllable and scalable, from small stand-alone spaces to multiple rooms or areas, to whole floors, entire buildings, and even whole campuses. With digitally addressable ballasts, light fixtures can be directly networked with time clocks and occupancy sensors—not to mention daylight sensors, wall controls, handheld remote lighting controls, window shades, building management systems, and each other. Since they are digital, they can be easily reconfigured, so that as spaces change, lights can easily be regrouped into different zones or to work with different sensors without rewiring. Fluorescent luminaires that use these ballasts allow for light levels from 100% down to 1% of full light output, which allows for many energy-savings and productivity-increasing strategies. What's more, digitally addressable ballasts allow users to monitor and report on the energy usage and functionality of each luminaire. In the not-too-distant future, the luminaire will be the lighting equivalent of a VAV box in an HVAC system. Direct digital control (DDC) systems will control both the amount and quality of the light.

- Recircuit or rezone lighting to allow personnel to only turn on zones based on use rather than operating the entire lighting system.
- Personal control: Install personal lighting controls so individual occupants can vary the light levels within their spaces. Personal light control allows users to control general lighting directly over their workstations. The ability to vary lights to the appropriate level for the job at hand can improve productivity and reduce eyestrain and glare while saving energy. In fact, research by lighting expert Peter Boyce showed that "people with dimming control reported higher ratings of lighting quality, overall environmental satisfaction, and self-rated productivity" (Boyce et al. 1999). Lighting energy savings from personal control is usually at least 10%.
- High-end trim/tuning: Lighting electricity usage can be reduced by 20% or more through high-end trim, which sets the maximum light level for each space. For example, the human eye can barely distinguish between a 100% light level and an 80% light level—but setting lights to 80% reduces energy use by about 20%. Light-level tuning sets the appropriate target level for each space, which is lower than the high-end trim level. So even when you employ high-end trim, many occupants prefer lower light levels to minimize glare on computer screens.
- Daylight harvesting: Daylight harvesting automatically dims electric lights when enough daylight is present. A daylight harvesting system can typically save an additional 10% to 60% in lighting electricity costs in buildings with many windows or skylights. This is addressed separately later in this section.
- Consider installation of lighting systems that facilitate load shed requests from the electric utility or energy aggregator.
- Evaluate turning emergency lighting OFF or to a lower level when a building or portion of a building is completely unoccupied without sacrificing safety requirements.

Exterior Lighting. The following are options to control and reduce exterior lighting usage:

- Use automatic controls that can reduce outdoor lighting levels or turn them OFF when either sufficient daylight is available or when not needed.
- Reduce power levels or turn exterior signage OFF when appropriate.
- When selecting new outdoor luminaires, consider the amount of backlight, uplight, and glare delivered by each luminaire type to improve functionality and minimize environmental impact. See the Illuminating Engineering Society TM-15-2011 (IES 2011c).
- Astronomical time clock scheduling: Scheduling automatically dims or turns lights OFF at certain times of the day. Few buildings operate on 24-hour schedules, and many are empty during the overnight and weekend hours. Astronomical time clocks can be used to provide a building lighting sweep at night so that lights are turned OFF or set to a low-dimmed level at certain times, saving energy and preventing light pollution. Astronomical time clocks are preferable to standard time-of-day time clocks because they can automatically adjust lighting based on astronomical events such as sunrise or sunset, ensuring lights are not wasting energy when they don't need to be ON. Scheduling can reduce lighting costs by 10% to 35%.

Figure 5-4 shows typical energy savings from using some of the light control strategies:

Use Daylighting. Take advantage of daylight to illuminate existing buildings. Lighting in daylighted areas should be separately controlled, preferably with automatic daylight sensors that automatically dim electric lights when enough daylight is present. In buildings with many windows or skylights, a light control system that uses daylight harvesting can save an additional 10% to 60% in lighting electricity costs. To ensure maximum savings, daylight harvesting light controls should be partnered with dimming ballasts, not switching ballasts. Daylight harvesting with stepped-switching systems can only step the lights at predetermined levels (i.e., from 100% to 50%), which means that there are no energy savings until a major threshold is crossed. With continuous dimming ballasts, daylight sensors start to dim the lights as soon as daylight is sensed in the space, thereby starting to save energy immediately. Also, building managers should be mindful of solar heat gain and glare which can make occupants uncomfortable. They should provide user-operable blinds, shades, or curtains on all windows to give occupant shielding from potentially hot (or cold) window surfaces and reduce glare.

Controllable window treatments (i.e., window shades) serve a dual purpose—to let daylight in and to keep excess heat and cold out. For total control of the visual environment, window treatments can automatically open and close at different times of the day in order to harvest daylight and reduce HVAC costs. This operation can reduce HVAC costs by as much as 30%.

Systems can be added that increase the effectiveness of daylighting. These can include the addition of top lighting through skylights and tubular daylighting devices (see analogous new requirements for minimum top lighting in ASHRAE Standards 90.1 [2010c] and 189.1 [2009a]), addition of light shelves to redirect light, and replacement of the top pane in split-level glazing with higher VT glass (perhaps in combination with a light shelf). These systems can be costly and are usually considered only when major building or space rehabilitation is under way.

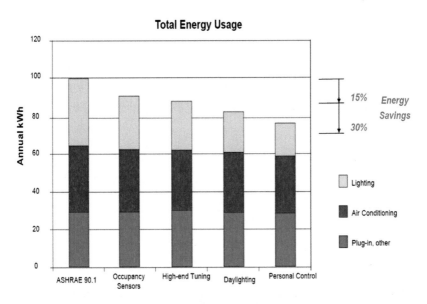

Figure 5-4. Typical energy savings from lighting control strategies.

MOTORS

Motors are found in most systems in commercial buildings, used in applications such as pumping, fans, and conveyance. Building owners should always specify NEMA (National Electrical Manufacturers Association) premium efficiency motors for all new installations and failed or end-of-life motor replacements. In such cases, any first-cost increment will be paid back through energy cost savings in a few years or less. For retrofit applications, it is necessary to evaluate the energy savings compared to the cost of purchasing and installing new, premium efficiency motors. Table 5-10 shows efficiencies for premium efficiency motors according to NEMA Standards (NEMA 2010), which can also be accessed on the NEMA Web site: www.nema.org.

MotorMaster+ (DOE 2010) is a free online NEMA Premium efficiency motor selection and management software tool from the DOE's Office of Industrial Technology (OIT). Note that it is applicable to commercial, institutional and even multifamily facilities as well. It is a very valuable tool that can be used to evaluate the upgrade or change out existing motors, and employed to set up and run a motor management program for your building or multisite organization. The *Compare* feature of the software can quickly do very exacting analyses of the savings potential of replacements or repairs of single motors, or *Batch* analyses of all motors (above fractional horsepower) within an organization.

MotorMaster+ can also be used to set up an overall motor management program and track all your motors over time. This takes a bit more time but in any organization with more than a couple of dozen 1 HP+ motors it is worth the effort.

One of the real strengths of *MotorMaster+*, in addition to the analysis capabilities, is its extremely rich library of specifications (which include in many cases part-load efficiencies and even list prices) for over 20,000 motors. A free copy of the software can be downloaded from the DOE Web site (www1.eere.energy.gov).

When replacing or installing motors, it is important to verify proper motor shaft alignment and/or belt tension, because these will affect performance and overall efficiency.

Proper motor sizing can be an important consideration when purchasing and installing a motor. For motors greater than 1 hp, efficiency is

Table 5-10. Nominal Efficiencies For "NEMA Premium™" Induction Motors Rated 600 Volts Or Less (Random Wound)

hp	Open Drip-Proof			Totally Enclosed		
	6-pole	4-pole	2-pole	6-pole	4-pole	2-pole
1	82.5	85.5	77.0	82.5	85.5	77.0
1.5	86.5	86.5	84.0	87.5	86.5	84.0
2	87.5	86.5	85.5	88.5	86.5	85.5
3	88.8	89.5	85.5	89.5	89.5	86.5
5	89.5	89.5	86.5	89.5	89.5	88.5
7.5	90.2	91.0	88.5	91.0	91.7	89.5
10	91.7	91.7	89.5	91.0	91.7	90.2
15	91.7	93.0	90.2	91.7	92.4	91.0
20	92.4	93.0	91.0	91.7	93.0	91.0
25	93.0	93.6	91.7	93.0	93.6	91.7
30	93.6	94.1	91.7	93.0	93.6	91.7
40	94.1	94.1	92.4	94.1	94.1	92.4
50	94.1	94.5	93.0	94.1	94.5	93.0
60	94.5	95.0	93.6	94.5	95.0	93.6
75	94.5	95.0	93.6	94.5	95.4	93.6
100	95.0	95.4	93.6	95.0	95.4	94.1
125	95.0	95.4	94.1	95.0	95.4	95.0
150	95.4	95.8	94.1	95.8	95.8	95.0
200	95.4	95.8	95.0	95.8	96.2	95.4
250	95.4	95.8	95.0	95.8	96.2	95.8
300	95.4	95.8	95.4	95.8	96.2	95.8
350	95.4	95.8	95.4	95.8	96.2	95.8
400	95.8	95.8	95.8	95.8	96.2	95.8
450	96.2	96.2	95.8	95.8	96.2	95.8
500	96.2	96.2	95.8	95.8	96.2	95.8

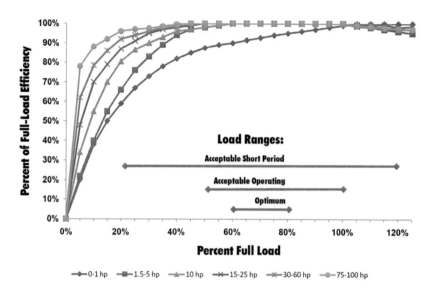

Figure 5-5. Part-load motor efficiency from full-load efficiency (DOE 2011a).

typically the highest when loaded between 50% and 100%. The typical relationship between motor efficiency and load is presented in the Figure 5-5 (DOE 2010). Determination or estimation of motor loading can be done using the methods presented in the DOE publication. If motor loading is not known, a good general rule-of-thumb would be to estimate it to be between 75% and 85%.

Savings Calculations. A simplified approach to estimating energy savings from a motor efficiency improvement is shown as follows:

$$kW = HP \times \%Load \times 0.746 \times \left[\frac{1}{\eta_o} - \frac{1}{\eta_{ee}}\right] \tag{17}$$

$$kWh = kW \times OpHrs$$

where:

HP	=	motor nameplate horsepower, hp
%Load	=	operating load (% of nameplate hp)
0.746	=	conversion from hp to kW

η_o =   efficiency of existing or standard motor, expressed as a decimal

η_{ee} =   efficiency of new energy efficient motor, expressed as a decimal

OpHrs   =   motor operating hours

CONTROLS

Building controls, particularly centralized building controls that are part of an EMCS, can perform a wide range of functions. Energy systems almost always have some type of control, even if only a switch on the wall. Improving the level of control of energy systems—lighting, heating, cooling, ventilation, and others—often provides some of the most impressive energy savings in buildings. In previous sections, potential savings from reducing operating hours (often the largest savings are obtained in this manner), adjusting temperature settings, and improving lighting control have been discussed. These types of energy improvements often are achieved through controls changes or improvements.

In smaller buildings, controls equipment tends to be simple and manual. As building size increases, investments in centralization of controls also tend to increase, since the centralization helps to make systems more effective. Diligent manual control in small buildings can achieve important energy savings, but as building size becomes larger, manual control becomes more difficult to accomplish due to the time involved. In commercial buildings, activities of the enterprise will usually crowd out manual energy management activities. A higher level of expertise is often needed to identify controls measures, so an energy professional will usually be needed to examine controls.

Critical Nature of Controls and Controls Fixes

Commissioning of existing building energy systems is receiving increased attention year after year, since important savings have been found for these types of activities. In a recent study, information on commissioning measures implemented under utility programs across the United States was collected from 11 utilities known to conduct existing building commissioning programs. In total, data on 122 commissioning projects and over 950 commissioning measures were obtained for this recent study: *A Study on Energy Savings and Measure Cost Effectiveness of Existing Building Commissioning* (PECI 2009). The measures were categorized into 44 measure types. EUI savings (kBtu/ft^2/yr) and simple payback were calculated for each measure type. Controls measures constituted most of the top savings measures and the majority (91%) of the 44 measure types had a simple payback of less than two years. Figure 5-6, controls savings, shows the simple payback periods and the range of EUI savings for several of the measures. Since most of these measures are controls measures, the importance of controls savings—and keeping controls working correctly—can be seen. Note that in the figure, the measures with higher EUI savings are to the left.

A list of abbreviations used in Figure 5-6 is given below:

A/C	air conditioning
CHWST	chilled-water supply temperature
CWST	condenser-water supply temperature
DCV	demand control ventilation
dP	differential pressure
DSP	ductstatic pressure
OA	outdoor air
occ	occupied
SAT	supply air temperature
VAV	variable air volume
VFD	variable-frequency drive

The results displayed here indicate the need to assure that HVAC and lighting controls are kept working properly, and checking control on a regular basis can be important to keeping EUIs in line with goals. If controls have not been checked for years, important savings may be possible. Figure 5-6 can be used as a priority list in approaching controls opportunities.

EMCS

An energy management and control system (EMCS) is an example of a distributed control system. The control system is a computerized, intelligent network of electronic devices, designed to monitor and control many systems in a building, including energy systems. Basic functionality of an EMCS keeps the building climate within a specified range, provides lighting based on an occupancy schedule, monitors system performance and device failures, and provides notifications to building engineering staff on equipment performance issues. Many other functions can also be provided.

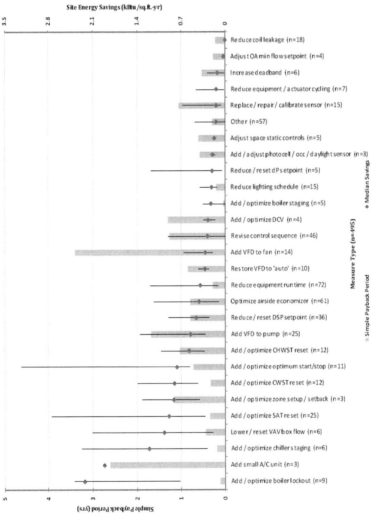

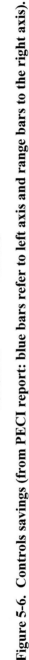

Figure 5-6. Controls savings (from PECI report: blue bars refer to left axis and range bars to the right axis).

System complexity remains an issue for EMCS, and although capabilities and complexities usually increase with building size, user knowledge and abilities tend to lag behind system capabilities. So although an EMCS can play an important role in controlling energy use and achieving energy savings, the complexities of both the EMCS itself and the control interactions with the energy systems mean that care is required to assure systems keep running efficiently. Regular tracking of EUIs and ECIs can indicate when controls issues may arise, and an EMCS can be used to help in this tracking.

COMMISSIONING AND RETROCOMMISSIONING

Commissioning is one of the most important energy efficiency measures that can be undertaken. The operation of building systems must be periodically or continually verified or they will drift from the intended state of operation, wasting energy. Commissioning is a process of ensuring that energy using systems in a building are functioning as intended. Retrocommissioning refers to returning the function of existing building systems to the as-designed state. Modern building systems are complex. It is important to verify that new systems function as intended. Once installed, the proper operation of these systems must be maintained. Energy savings from commissioning are typically 10% of the building energy use, but can be as much as 30%.

ASHRAE guidelines relating to commissioning include:

Guideline 0: The Commissioning Process (ASHRAE 2005)

Guideline 0.2P: The Commissioning Process for Existing Systems and Assemblies (ASHRAE 2008a)

Guideline 1.1: HVAC&R Technical Requirements for the Commissioning Process (ASHRAE 2007c)

Guideline 1.2P: The Commissioning Process for Existing HVAC&R Systems (ASHRAE 2006a)

Guideline 1.3P: Building Operation and Maintenance Training for the HVAC&R Commissioning Process (ASHRAE 2006b)

Guideline 1.4P: Systems Manual Preparation for the Commissioning Process (ASHRAE 2009b)

IES guideline for commissioning of lighting systems:

DG-29-11, *Commissioning Process Applied to Lighting and Control Systems* (IES 2011a)

Refining Financial Analysis And Setting Priorities 6

For small buildings and efficiency improvement measures with a payback period of under three years, simple payback is probably adequate to make decisions, For larger buildings and longer payback periods, more sophisticated financial analyses are advisable. Life-cycle costing (LCC) is used to evaluate the total cost of ownership of energy efficiency measures. LCC accounts for factors such as the time value of money, escalation of energy costs over time, annual maintenance costs, component replacement costs, and the useful life of the equipment. Other factors that may also be considered include temporary disruption of building operations. Major measures in schools are often implemented in the summer when the building can be empty. Measures that affect the entire building may require swing space and temporary movement of departments.

LIFE-CYCLE COSTING FOR MEASURE SELECTION PRIORITY

Quite often, the identification of energy efficient measures is a straightforward task; added insulation, T8 light changes, and efficient equipment upgrades are so common that they can be offered as viable choices before a site audit is even conducted. However, in most cases, the sum of the costs of attractive energy measures is greater than the available funds. A means to establish the worth of a particular project is required to appropriately allocate limited funding. In line with typical capital investment considerations, the future benefit of investment should outweigh the initial cost. A range of evaluation tools are available to discern the most valuable projects to fund, from simple payback to quantification of return on investment. However, the evaluation tool which yields the most comprehensive analysis is called *life-cycle cost analysis*.

When presented with a financial decision, most building operators and owners are familiar with the concept of a simple payback to determine a measure of worth. For this method, the total initial cost of a project is divided by the annual revenue (savings for EEMs) returned to yield an amount of years required to "pay back" the investment. While this can be used as a quick "back of the napkin" type of analysis to determine if further analysis is necessary, the reader is cautioned not to rely on this method as a final consideration, as it overlooks certain criteria that are critical for proper project comparison. All other costs (including financing) and the time value of money are excluded, quite often with detriment, from this type of evaluation. In addition, this method assumes that there is no benefit once payback is achieved. In other words, simple payback states that the life of the investment is irrelevant once the initial cost is recouped.

In the simple payback method, the scope of additional costs is not accurately represented. Initial costs are assumed to be the purchase price of new equipment and the monetary compensation of labor for its installation. Additional initial costs often overlooked include: disposal fees of preexisting equipment (including possible environmental costs), design fees of engineers and architects, and permitting fees. Also, recurring costs are experienced over the life of the investment that are not captured by simple payback. Annual expenses, like maintenance, or periodic (nonannual) replacement costs should be included in the complete analysis. Also, any value associated with equipment at the end of the life cycle, called *salvage value*, needs to be ascertained. Even the capital used to fund the project has a cost associated with it.

The cost of capital is a function of the method of financing and relevant tax rates. Debt and equity financing are the two main categories for fund resource. Debt is financed through loans and bonds which impart no ownership to the lender, while equity funds are collected from stocks and retained earnings that share ownership with the lender. Debt financing offers a tax incentive since the Federal government allows the borrower to deduct interest payments on debt from the company's gross income, which reduces their taxable income. This reduction in taxation would be beneficial to include in a complete project economic analysis. Additionally, depreciation, or the reduction of an asset's value, can be included to further reduce the taxable income. Conversely, tax credits that also lower

taxable income may be available specifically for energy-related investments for some end users.

The lifetime of an investment has concluded once the final cash flow
has occurred. Since the benefit of a capital investment is realized in the
future, the sensitivity of the buying power of money to time needs to be
considered for accurate investment scrutiny. Not only does LCC peer
beyond the payback period but it does so while accounting for the time
value of money. Two factors influence the worth of money in relation to
the future: interest and inflation. *Interest* is a return on money that is
loaned, while *inflation* is the reduction of the purchasing power of money.

With interest rate in mind, several approaches to quantify the worth of a
particular project, or measure, can be defined. All but one of these
approaches uses a figure called the *minimum attractive rate of return*
(MARR) as the interest rate. Debate surrounds the actual definition of the
MARR, but it is typically agreed to be greater than the cost of capital used
to fund the measure. In cases where funds are retrieved from a variety of
sources, a weighted average cost of capital is derived.

The most straightforward approach of LCC is the present worth, or
net present value (NPV), method. NPV evaluates various options by
relating all future cash disbursements and receipts into a lump sum
value in current dollars. Priority would be placed on the largest NPV.
Related to this technique is annual worth, which renders all cash flows
into a uniform series of annual payments. Beneficial investments are
revealed if the annual worth is greater than zero. These methods use a
MARR, which is often assumed and subjective to the company or decision maker. The MARR can be thought of as an interest rate that is used
to correlate future value of an investment to a worth in current dollar
value. Since the MARR is subjective, it can fluctuate if investment conditions change. When this occurs, the present or annual worth would
need to be recalculated to determine if a benefit can still be yielded by
the investment. At times, due to increased uncertainty of a specific and
constant MARR, an alternate method called the *internal rate of return*
(IRR) is a better choice for investment analysis. This method attempts
to minimize the influence of the MARR subjectivity by deriving a rate
of return that is specific, or internal, to the particular project under scrutiny. In this case, an interest rate is determined which yields a net present value of zero (which indicates a neutral bias). An attractive
investment is realized when the IRR is greater than the MARR. The

attractiveness of the investment can be quickly reevaluated without recalculating the entire investment benefit of the project if, for some reason, the MARR changes. In practice, this calculation is an iteration that can be cumbersome to generate since it is found through trial and error.

In conclusion, life-cycle cost analysis incorporates a wider spectrum of data to yield a more accurate evaluation of potential investments. From initial costs to recurring costs and even the cost of the funding capital, LCC presents a dynamic assessment of the performance of a project. Although many businesses use at least the ROI method to determine most investment decisions, energy decisions always seem to be measured by simple payback. Adopting an LCC approach for all investment decisions would finally place energy considerations on equal footing with other demands for capital funding.

SIMPLE PAYBACK VS. LIFE-CYCLE COST ANALYSIS

Many energy efficiency decisions are made on the basis of simple payback period. Since no other business decision is made using this *payback* parameter, building owners and managers are not evaluating these improvements on an equal footing against other investments. If an owner of five properties becomes interested in another building that comes onto the market, he or she doesn't think, "What is the payback on that purchase?" If the manager of an office management firm has a major tenant leave two floors of one of the firm's sites and he decides to repaint and carpet the vacant spaces, he does not ask "What will be the payback on those upgrades?" No, he wants to know for how much more the space can be rented, or how that affects the building's cash flow. The owner deciding on purchasing that additional site wants to know the net present value of that investment, or what the ROI or IRR is for that capital expenditure. By utilizing the *payback* criteria, we have segregated out energy improvements from all other competing options that a building ownership/management organization has for its limited pool of capital. Energy efficiency improvements, which many times might have a better economic impact on the building or company, often do not compete for those limited funds. Hence, cost-effective projects are not properly evaluated and do not get implemented.

For these reasons, even if a full life-cycle costing does not get undertaken, at the very least a simple ROI should be calculated for evaluation against other alternative investments. While a simple ROI (%) is nothing more than the inverse of a simple payback, at least it gets the decision-maker thinking in business terms, putting these (EEM) alternatives on the table and giving them a chance.

GOVERNMENT VS. PRIVATE OWNERS AND MANAGERS

Commercial business enterprises often have a different perspective than governments/municipalities when evaluating the financials of an investment. Private owners and managers generally focus on near-term payback of investments and comparative return on investment versus other opportunities. The manufacturing and industrial sectors in the United States commonly seek only projects with simple paybacks of less than three years. Conversely, government and municipal decision-makers often accept much longer simple paybacks or evaluate life-cycle costs when selecting a project investment choice.

The availability of capital funds also plays a significant role in energy efficiency investment decisions. In the past, and likely in the foreseeable future, government and municipal clients have severely limited available capital for investment in energy efficiency. This situation, combined with the potential to secure tax-free financing, makes ESCO projects (energy performance contracts and guaranteed savings projects) a viable and attractive procurement approach. Many government and municipal buildings have paid so little attention to energy efficiency and upgrades that they are a good fit for a comprehensive project encompassing numerous mechanical, electrical, and building shell systems.

COST OF CONSERVED ENERGY AS A DECISION-MAKING PARAMETER

The cost of conserved energy (CCE) is a measure that can be used to judge the cost-effectiveness of an energy efficiency investment. CCE is based on a pair of ideas. The first of these is that energy "freed up" by efficiency improvements can be considered as a resource. Secondly, the cost of the saved, or conserved, energy can be compared to the cost of the energy such as electricity, natural gas, or oil, we would otherwise be purchasing to satisfy a load.

The cost of conserved energy must consider all financial factors that impact the investment. If money is borrowed, the cost of interest must be considered. In the case of for-profit companies, energy conservation will increase profits. Taxes will need to be paid unless the savings are invested in other deductible expenses.

Consider the following example. Let us suppose that *XYZ Property Management* consumes four million kWh per year in its largest office park. One of the vice presidents presents plans to build another building at that campus that would require 300,000 kWh per year in additional energy. If XYZ Property Management accepts the vice president's proposal, it has a choice: it can either purchase another 300,000 kWh per year, or it can try to find energy efficiency and conservation opportunities in the complex to reduce consumption by 300,000 per year, thereby freeing up those kWh for the new building.

An engineer for XYZ Property Management uncovers that an investment in a new central chiller plant will reduce energy consumption by 300,000 kWh of electricity per year. Working with the chief financial officer's staff, they compute that by spreading the cost of the investment over the useful life of the new equipment it will amount to $13,500 per year. They construe the analysis to indicate that it will cost the company $13,500 per year to save those 300,000 kWh, or 4.5 cents per kWh. That $0.045/kWh is the CCE.

The CCE can be contrasted against the cost of buying additional energy, electricity in this case. If the current price of electricity is in the range of $0.12 to $0.14 per kWh, assuming the company's effective tax rate is 50%, the after-tax cost to the firm is $0.06 to $0.07/kWh. The evaluation clearly indicates that the company shall reduce costs, and in turn maximize profits, by investing in efficiency instead of purchasing more electricity. That is because conserved energy at 4.5 cents per kWh is cheaper than purchased electricity at 6 to 7 cents per kWh.

The CCE is a powerful tool for professionals to use in the analysis of, and demonstration to, management of the value and cost-effectiveness for proposed EEM investments. This parameter can be computed for any investment the reduces the amount of energy used in a building.

Basically, CCE is computed as:

CCE = Effective Annual Project Cost ($ per yr) / Annual Energy Saved(1)

(Envest 1985). The effective annual project cost is estimated by first computing the present value of the nonenergy costs (and, where applicable, benefits) of owning and operating the EEM, then spreading the present value over the project life.

LIFE EXPECTANCY

The service life of energy measures is one parameter needed to conduct life-cycle cost analyses.

In the state of Washington, public agencies are responsible for ensuring that energy conservation and renewable energy systems are considered in the design phase of major facilities. This consideration involves selection of low life-cycle cost alternatives, based on life-cycle cost analysis. As part of the guidelines for conducting life-cycle cost analyses, Washington provides information on energy measure equipment service lives. This information is provided for reference in Table 6-1: Equipment Service Life 1 (Table 4.6 of www.ga.wa.gov/eas/elcca/simulation.html [GA 2006]) (McRae 1987; ASHRAE 1995). Additional information is available in the 2011 *ASHRAE Handbook—HVAC Applications*, Chapter 36 (ASHRAE 2011b).

ASHRAE has developed a free online database that provides information on equipment service life and annual maintenance costs for a variety of building types and HVAC systems. The database can be accessed at www.ashrae.org/database (ASHRAE 2010d). The database contains more than 300 building types and more than 38,000 pieces of equipment with service life data. It allows users to access up-to-date information to determine a range of statistical values for equipment owning and operating costs. This online tool does require a specific data query to obtain extensive information on particular classes of equipment, but users should find it valuable.

This database is meant to improve on the information in previous versions of the equipment life expectancy table for Table 6-1 (BPA 1987; ASHRAE 1995). The database is actually an ongoing project (ASHRAE Research Project 1237 [Abramson 2005]) to provide such data for public use. As part of the project, users are encouraged to contribute their own service life and maintenance cost data to help further the utility of this tool. Over a period of time, user input is hoped to provide sufficient service life and maintenance cost data to allow the comparative analysis of various HVAC systems types in a broad variety of applications. Data can be entered by logging into the database and registering.

Information from the database is also used to update *ASHRAE Handbook—HVAC Applications*, Chapter 37, Owning and Operating Costs (2011c). This chapter contains median HVAC equipment service life data

Table 6-1. Equipment Service Life (in years)

Heat Pumps	Pumps	Package Chillers
Commercial air-to-air (15)	Base mounted (20)	Absorption (23)
Commercial water-to-air (19)	Condensate (15)	Centrifugal (23)
	Pipe mounted (10)	Reciprocating (20)
Residential air-to-air (15)	Sump and well (10)	Scroll or screw (20)

HVAC	Air Terminals	Refrigeration
		Automatic cleaning system for condenser tubes (15)
Air Conditioners	Diffusers, grilles, and registers (27)	
Commercial through-the-wall (15)	Induction and fan-coil units (20)	Condenser floating head pressure control (10)
Computer room (15)	Low-leakage damper (9)	Hot-gas bypass defrost (10)
Residential single or split-package (15)	VAV and double-duct boxes (20)	Polyethylene strip curtain (3)
Roof-top multizone (15)	Variable inlet vane dampers (20)	Refrigeration case cover (11)
Roof-top single-zone (15)		
Water-cooled package (15)	Duct work (30)	Unequal parallel refrigeration (14)
Window unit (10)	Air-side economizer (10)	

Condensers	Coils	Radiant Heaters
Air-cooled (20)	DX, water, or steam (20)	Electric or gas (10)
Evaporative (20)	Electric (15)	Hot-water or steam (25)

Controls	Boilers, Hot Water (Steam)	Cooling Towers
		Ceramic or fiberglass reinforced plastic (34)
Computer-logic EMS (13)	Cast iron (30)	Galvanized metal (20)
Deadband thermostat (13)	Electric (15)	Variable-pitch cooling tower fan (13)
Electric controls (16)	Steel fire-tube (25)	
Electronic controls (15)	Steel water-tube (30)	Wood (20)
Pneumatic controls (20)	Burners for boilers (21)	Chiller-strainer cycle economizer (15)
Time clocks (10)	Steam traps (7)	Waterside economizer (11)

Table 6-1. Equipment Service Life (in years) *(continued)*

Unit Heaters	Thermal Energy Storage Systems	Valve Actuators
Electric or gas (13) Hot-water or steam (20)	Ice (19) Water (20)	Hydraulic (15) Pneumatic (20) Self contained (10)
Motors and Drives	**Lighting**	**Heat Recovery**
High-efficiency electric motor (17) Motor starters (17) Standard electric motor (15) Variable-speed DC motor (18) Variable-speed drive–belt type (10) Variable-speed drive–solid state (15)	Compact fluorescent: detachable ballast (12) Dimming systems (20) Ballast: all types (12) Lighting fixtures: fluorescent, HID, etc. (20) Motion sensor (10) ON-OFF switching (7) (Note: For lamps, use tested lamp life)	Heat recovery from refrigeration condensers (11) Plate-type/heat-pipe recovery system (14) Rotary-type heat recovery system (11) Makeup air unit for exhaust hood (10)
Furnaces	**Heat Exchangers**	**Reciprocating**
Gas- or oil-fired (18)	Shell and tube (24)	Compressors (20) Engines (20)
Domestic Hot Water	**Fans**	**Building Envelope**
Heat-pump water heater (10) Point-of-use water heater (12) Solar water heater (15)	Axial (20) Centrifugal (25) High-inlet/low-discharge-type air destratification (15) Paddle-type air destratification (10) Propeller (15) Ventilating roof-mounted (20)	Air curtain (10) Blanket insulation (24) Molded insulation (20) Solar shade film (7) Tinted and reflective coating (14)
Other		
Electric Transformers (30)	Air Washers (17)	Steam Turbines (30)

from the first 163 commercial buildings that were used to seed the database, with updates to come as the database grows.

MAINTENANCE COSTS

In many cases, potential reductions in maintenance costs from installing new energy efficiency measures can be an important factor in deciding how to prioritize measures or whether to install new ones or not. For example, poor water quality can have major impacts on boiler or chiller efficiency and equipment lives, and improving water quality might be a major factor in extending equipment life and improving equipment efficiency. In order to determine what impacts specific categories of maintenance costs are having on total building operating costs, each building must have some means of tracking maintenance cost data by energy system type, or overall if costs are low. In practice, building owners and managers often track such costs by maintenance *craft,* which may include categories similar to energy system types.

Some examples of craft type for energy systems are:

- HVAC
- Electrical
- Energy management/security systems
- Other mechanical
- Elevators

In some cases, more detail may be needed on HVAC or electrical maintenance costs in order to reasonably conduct an economic analysis of potential energy efficiency measures. Maintenance activities might be handled in-house or contracted, and both types of costs must be considered, possibly in different ways. Fortunately, in most cases maintenance costs are not a major factor affecting decisions about whether to install energy efficiency measures or what priority to give to measures. But if maintenance costs are high for energy-using systems, care must be given to understand how to treat these costs in an economic analysis of potential efficiency measures.

For comparison of a specific building or portfolio of buildings maintenance costs with others, *The Whitestone Facility Maintenance and Repair Cost Reference* provides building maintenance & repair cost statistics (www.whitestoneresearch.com) (Abate 2010).

BOMA is the Building Owners and Managers Association International. The BOMA Experience Exchange Report (www.boma.org/resources/benchmarking/Pages/default.aspx [BOMA 2010]) is developed on a regular basis based on input from building owners and managers about their own buildings. The 2010 report tracks income and operating expenses from 4200 buildings and 110 locations across North America, including:

- Office rents
- Retail and other rental income
- Telecommunication and wire access income
- Real estate taxes
- Energy and other utilities costs
- Repairs and maintenance
- Cleaning
- Administrative costs
- Security
- Roads and grounds

The BOMA data on energy, utilities, and repairs and maintenance can help compare to large office building costs in these areas. The Experience Exchange Report is a subscription document, with a subscription lasting until the next version of the report is published. The costs shown in Table 6-2 are for the 2010 subscription.

Table 6-2. BOMA Experience Exchange Report Cost

	Survey Participants	**All Others**
Single market	$95	$129
Each additional market	$20	$49
Full access	$195	$279

The ASHRAE database provides information on total maintenance costs and as of 2010 had data on 267 buildings from around the country (2010d). The ASHRAE data can be accessed for free: www.ashrae.org/database. The ASHRAE data cover several building types and have data from buildings in many states. The major breakdowns provided are by:

- Building function (primary space type)
- Building size
- Building age
- Building height (stories)
- BOMA class (A, B, or C)
- Location (downtown, suburban, rural)
- State
- Census region
- All buildings

As an example, the current breakout (ASHRAE 2010d) by building type is shown in Table 6-3: Maintenance 1.

The standard deviation (Std Dev) should be viewed with caution, as some of the data are heavily skewed (i.e., compare the max value to the mean), and the standard deviation in those cases only provides an indication of how heavily skewed the outliers are. Similarly, when the mean is much different than the median, data distribution issues are indicated. Maintenance costs of zero reiterate the issue that a wide range of activities and issues affect buildings data, and extreme values must be considered carefully in all building data distributions.

PRIORITIZED RESULTS AND IMPLEMENTATION STRATEGY

The prioritized analysis results of potential energy efficiency measures described previously should allow sorting and grouping of potential measures to compare against desired progress toward efficiency goals established at the beginning of the overall process. Energy and cost savings and resources required to implement are typical factors used to sort and group measures. Required resources can include both financial and human resources.

Table 6-3. Maintenance 1

Building Function	No. of Buildings	Mean, $/ft^2	Median, $/ft^2	Std Dev, $/ft^2	Max, $/ft^2	Min, $/ft^2
Office	149	$0.45	$0.36	$0.59	$7.05	$0.00
Medical	1	$0.78	$0.78	n/a	$0.78	$0.78
Lodging	2	$0.69	$0.69	$0.72	$1.19	$0.18
Manufacturing	0	n/a	n/a	n/a	n/a	n/a
Retail	15	$0.11	$0.11	$0.10	$0.41	$0.00
Residential: multifamily	11	$0.35	$0.14	$0.36	$1.18	$0.05
Dormitory/ Barracks	5	$0.48	$0.27	$0.68	$1.66	$0.01
School (K–12)	40	$0.11	$0.08	$0.16	$0.91	$0.01
School (college/ university)	2	$0.09	$0.09	$0.04	$0.12	$0.06
Laboratory	3	$0.67	$0.84	$0.55	$1.11	$0.05
Restaurant	1	$0.22	$0.22	n/a	$0.22	$0.22
Public assembly	0	n/a	n/a	n/a	n/a	n/a
Warehouse/ Storage	0	n/a	n/a	n/a	n/a	n/a
Other	38	$0.14	$0.07	$0.20	$1.07	$0.00

Once potential measures are sorted and grouped and compared against goals, the implementation strategy can be developed and a time line for planned progress toward meeting goals can be established.

SELECTING THE NEAR-TERM LEVEL OF IMPROVEMENT OR LEVEL OF RENOVATION

All buildings should be operated under a long-term energy efficiency improvement plan. Some elements of the plan may not be implemented for many years, depending upon the life cycle of building systems and the finances of the building. However, having a long-term plan in place prevents building owners from making a wrong decision when unanticipated circumstances arise. For example, installing VSDs on circulation pumps may be a midterm improvement. If the motor starter fails, should the starter be replaced by another now and then replaced by a VSD four years from now? Suppose ten years of efficiency improvements have been designed for implementation, culminating in replacing two large boilers with four modular condensing boilers of smaller total capacity. If a boiler fails unexpectedly, should it be replaced in kind, or with modular boilers? With a plan in place these decisions can be arrived at quickly. Without a plan, the replacement equipment is likely to be whatever is readily available.

What occurs in the near-term depends on the financial criteria of the owner, the age of building systems, timing of renovations, occupant turnover, and the energy efficiency of the existing building. Integration of the building energy efficiency plan with the overall financial plan will ensure that EEMs are part of the overall investment process for the building. Typically EEMs with the largest ROI are implemented first. However, measures that reduce load can result in smaller pumps, fans boiler and chillers.

Energy or carbon reduction targets are set relative to some measure of performance in the population or some historic level of performance. Once the energy performance of a building is established, such as an ENERGY STAR® score of 65 or 165,700 Btus/ft^2/yr, energy targets can be set based on the goals of the company and asset strategy. Based on the data sources discussed previously, examples of various energy targets are:

- A 10%, 20%, 30% or more improvement beyond historical performance
- Target the buildings with the highest energy use in a portfolio of buildings
- Qualify the building each year for the ENERGY STAR label (a score of 75 or more)
- Pursue LEED for Existing Buildings: Operations & Maintenance certification (a minimum ENERGY STAR score of 69 or 19th percentile of comparable buildings is required). (USGBC 2008)

ASHRAE has *Advanced Energy Design Guides* that can be followed. These reference documents are free and available for download from www.ashrae.org/publications/page/1604 (ASHRAE 2011a). For large buildings, or those not covered by an *Advanced Energy Guide*, the building's anticipated energy performance would be modeled to ANSI/ASHRAE/IES Standard 90.1 (2010c).

If the building is undergoing a major rehabilitation, see the next chapter.

Once financial analysis is completed, implementation can begin. Of course, with a multiyear plan it will be necessary to update analyses as conditions change.

Putting The Process Together 7

The previous chapters deal with specific topics, such as measuring building performance and implementing specific energy efficiency measures. The focus of this chapter is the overall process, as outlined in Figure 7–1. Building energy efficiency is achieved and maintained by making it an integral part of the overall building operations.

Improving energy efficiency is a continuous process. Successful execution requires:

- Measuring energy performance and setting efficiency goals
- Understanding how energy is used in the building
- Identifying and analyzing energy efficiency measures and savings
- Refining the financial calculations and setting measure priorities
- Preparing a plan for implementing
- Measuring priorities
- Installing or implementing the efficiency measures
- Commissioning energy efficiency measures
- Verifying savings achieved and progress toward goals
- Reporting success and continued tracking of performance
- Refining the process if additional improvements are desired

FULL RENOVATION DECISION

If the decision has been made to fully renovate the building, energy efficiency improvements should be approached in a similar manner to new construction as the building will need to meet or exceed current codes. Since all or most of the building systems are being replaced, this is an opportunity to dramatically improve energy efficiency and to

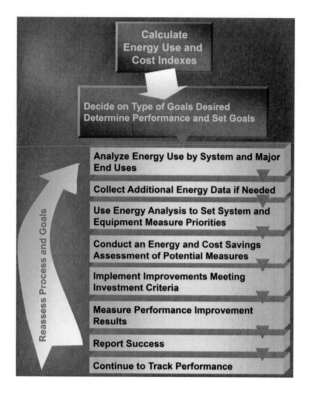

Figure 7-1. Energy efficiency process flow diagram.

address inherent deficiencies in the design/configuration of the existing systems. For many small- and medium-sized buildings, the *Advanced Energy Guides* can be used to develop EEMs.

For large buildings, or those not covered by an *Advanced Energy Guide*, the building's anticipated energy performance would be modeled to ANSI/ASHRAE/IES Standard 90.1 (ASHRAE 2010c). ENERGY STAR has a program called *Target Finder* that can be used to see how a building's modeled energy consumption ranks within ENERGY STAR (EPA 2011c).

Many utilities and state agencies offer free or cost-shared technical assistance to help identify and evaluate energy efficient system and equipment choice alternatives. Incentives are also often available to help offset the initial cost of implementation.

MAJOR PROCESS STEPS AND MANAGING
THE PROCESS

Managing building energy efficiency is a continual process. Building systems age and drift out of tune. Occupancy and internal loads change. Technology improvements result in the availability of more efficient building systems. Energy codes become more stringent and the EUI of the existing building population decreases.

Keeping a building current requires planning. Goals are set, resources mobilized, EEMs are selected and implemented, success is measured and then the process and goals are reassessed.

One key aspect of managing building energy efficiency is availability and long-term tracking of building energy use. Historical records of energy use, occupancy, etc. are key data to be used in evaluation of overall building efficiency.

Energy efficiency cannot be implemented in a vacuum. It must be accomplished in consort with other factors. These are beyond the scope of this handbook but must be mentioned. The goal is to improve energy efficiency, but not at the expense of occupant comfort or safety. Environmental conditions should be maintained and factors such as acoustics and indoor air quality must be considered.

Changes to a building can have unintended consequences. This doesn't mean take no action. Rather, it means ask the right questions before implementing measures and verify proper operation once the measures are installed. *Legionella* is a classic example; warm water in a dark place can grow bacteria. Some modern electronics are sensitive to power quality. Variable-frequency drives can impact power quality. The power supplies in personal computers and the transformers in compact fluorescent lamps can impact power quality as well. Another example is cool roofs. In most climates, cool roofs save energy by reflecting heat and improving the performance of the building insulation. However, if the insulation under a cool roof gets wet, it will not heat up enough in summer to dry out. A poor vapor barrier or a leak in the roof membrane can result in mold. But this can occur with a standard roof as well. Building economizer cycles only save energy when they are working properly. Many older buildings had two position dampers: open for occupied and closed for unoccupied. With an economizer cycle, a modulating damper is required. A properly installed and maintained economizer cycle will close the outdoor air damper when the building is in unoccupied mode.

But many systems, especially in smaller buildings, are not properly installed and maintained and do not fully close. These systems waste more energy heating and cooling outdoor air when no one is in the building than they save in economizer mode. Finally, the trend in building systems is to utilize ever more sophisticated controls. They must be properly programmed and maintained to achieve their full efficiency potential. The owner must ensure, before purchase, that different systems (e.g., the main building energy management system and the chiller control system) can communicate with one another. Building operators must ensure that when they override a setpoint, the controls will return to their former state after a period of time rather than remain overridden.

Define Commitment and Goals. An energy efficiency program begins with the commitment of the building owners. The owners must buy in to the concept that building energy efficiency is a desirable goal and will reduce the cost of doing business. Then they must be willing to commit the resources necessary to achieve that goal. These resources involve staff and capital:

- Energy awareness and energy efficiency must be incorporated into the manner in which the building is managed and operated.
- Financial resources must be put in place to ensure that energy efficiency measures can be implemented over time. Continual training of building staff is key to ensure the building systems are being operated in an efficient manner.

Both short- and long-term goals must be set and incorporated into a plan.

Define Initial Tasks. For large buildings, establish an Energy Management Team. A successful team should consist of building operations staff, finance, and management/executive level individuals.

Set some short-term goals. Immediate successes help to get building occupants involved.

Acquire and Align Resources to Tasks. Appoint an energy manager. One person should be ultimately responsible for and take ownership of the energy efficiency plan.

Obtain Buy-In of Staff. The success of an energy efficiency plan is dependent upon the building operating staff. They should be involved in the development of the plan so that they can take pride and ownership in the eventual success.

Hire an external energy auditor or energy efficiency consultant once the internal resources are in place to act on the results of the energy audit. Otherwise, the energy audit will be merely a report that sits on a shelf. Ensure that the results of the energy audit be delivered in spreadsheet format such that energy and implementation costs can be updated over time.

Involve Operations Staff in the Effort/Solution. The building operations and maintenance staff can be one, if not the most, critical factors in a building's energy use. It is crucial that they be made part of the official or unofficial energy team. We may one day live in a world where buildings run themselves and systems have so many monitoring points and so much intelligence that they control and self correct for all problems. We are not, however, at that point. Hence, a good operator can make or break the energy budget.

This staff should be trained on the equipment and controls utilized at the site. There are many good resources currently available for doing so, such as publicly offered courses (including some that result in certification, though that is not critical), Web-based, manufacturer-sponsored training, and union or trade association offerings. It is important that management encourages and supports such activities.

A proactive program of checking systems operations, rather then just reacting to "problem calls" should be initiated. Where possible, computer or Web-based monitoring systems that the staff has easy access to should be made available to assist in this activity. In absence of such, handheld devices (even those as simple as digital or infrared thermometers) and manual log sheets should be used. While this might appear to take away from more immediately pressing issues, time must be allotted for this task if the significant potential returns are to be reaped.

Operators need be brought into, and made to be felt a part of, the building or energy management team. They must often know the building better than anyone; if they feel a part of the solution you can leverage this extensive information base. They should be encouraged to develop and share in discussions about both small and large problems as well as potential renovation plans. Where possible, find ways to reward their efforts, as this will sow even greater returns.

Experience has shown that well-trained and motivated operator and maintenance personnel can not only significantly cut a building's energy budget, but makes for more comfortable conditions, reduces repair expenditures, and often increases the useful life of existing equipment.

FINANCING STRATEGIES

Major energy efficiency improvements can require substantial capital investment. Some of this capital should be built into the long-term plan for the building to cover building system replacement over the life cycle. Even in that instance, there is an incremental cost differential between replacing equipment in-kind, or to meet current codes, and replacing it with more efficient equipment. Incentives of various kinds are available in most locales to assist with the cost of the energy efficiency improvements.

Tax Credits. Tax credits are sometimes available through the federal or state government. These credits can be used to offset the cost of energy efficiency measures. The building owner should check the DOE Web site (www1.eere.energy.gov) (EERE 2011b) for federal tax credits and the state taxing authority for these credits. The building's accountant is another source for this information.

State or Utility Incentives. The state government or the local utility likely has programs that offer energy efficiency incentives.

Loans or Building Refinancing. In some jurisdictions, low-interest loans are available for the financing of energy efficiency improvements. These are usually offered through the local utility or state government.

As noted previously, DSIRE data base (NCSU 2011) is a comprehensive source of information on state, local, utility and federal incentives, policies, and programs that promote renewable energy and energy efficiency, but can indicate where some help might be found on energy efficiency efforts (www.dsireusa.org).

ESCOs. Many energy service companies will provide financing for energy efficiency improvements. Off-balance sheet financing can be attractive. However, it usually will cost more than owner financing.

Measure Performance Relative to Goals. Determine the building EUI and compare it to peer buildings. This will provide an indication of the energy conservation potential of the building. Goals can be set as a percentage energy efficiency improvement against the existing building baseline, against peer buildings, and ultimately as a percent improvement over the current code.

Analyze Energy Use and Efficiency Opportunities and Possibly Refine Goals. Meet with the energy auditor and develop a work scope for the energy audit.

Once the energy audit has been completed, it should be reviewed and discussed for incorporation into the energy efficiency plan.

Prioritize Opportunities and Refine Energy Savings. Meet with the energy auditor to prioritize energy efficiency opportunities and refine energy savings based upon the final list of measures. Prioritization may be based upon building operational needs, economic payback, available incentives or tax credits, and/or other metrics as appropriate.

OM&M and rapid payback measures provide immediate savings and create interest in the process.

Envelope measures may not be the most cost-effective, but their implementation can reduce the size of replacement system and plant equipment producing both cost and energy savings for those systems.

FACTORS INFLUENCING IMPLEMENTATION TIMING

Implementation of some measures will be disruptive to building operations, or must be coordinated with the life cycle of building equipment. Factors that should be considered include:

- Seasonal considerations such as implementing cooling improvement during the heating season
- Scheduling considerations—schools implement major improvements when classes are not in session
- Some measures must be implemented after hours or on weekends
- Creation of swing space for some alternatives
- Life cycle of existing equipment
- Controls—control issues are a prevalent problem and pitfalls should be discussed

Establish Expected Energy and Financial Savings Results. The energy-savings calculations prepared by the energy auditor are an estimate of savings. Results may vary somewhat, so the estimates must be verified and compared to expected improvements. Of course, bear in mind the relative cost of verification compared to the savings value itself. As appropriate, verify the more significant and more uncertain energy-savings measures. Large savings, or lack thereof, should show up in the monthly utility bills. But these will not indicate which measures are performing and which are not. Cost-effective M&V is accomplished by reviewing the energy savings estimates and verifying specific parameters in the savings calculations. Efficiency improvements in a steady load, such as an exit sign, can be verified by before/after one-time kW measurements. For loads that are steady but intermittent, the runtime must be verified. The length of time required to measure runtime will depend upon whether the runtime varies by season. Finally, variable loads will require measurement of consumption, (kWh or therms/h), not just runtime to determine savings. If the load varies with outdoor air temperature, measurements may be required for the whole heating and/or cooling season. If the building has an EMS, it often has the capability to trend variables and can be used for M&V. Many EMS systems lack sufficient memory for this task, but adding memory is not expensive.

Implement and Commission Improvements. Establish a commissioning plan for the measures. Commissioning is critical. The measures are being implemented to improve energy efficiency and commissioning provides assurance that those measures are working.

Once the commissioning plan is in place, the measures should be designed, implemented, and commissioned per the energy efficiency plan. The most successful commissioning plans are those that are developed and initiated as equipment retrofits are being specified and/or designed. Initiating commissioning as an afterthought after implementation limits the commissioning agent's ability to ensure the design and specification will meet the owner's requirements. For larger investments, use third-party commissioning agents.

Measure Progress towards Goals. The monthly energy bills (both cost and use) should be entered into a spreadsheet on a monthly basis. The building EUI can be tracked and updated monthly, although periodic adjustments for weather and occupancy may be required. Continuous tracking of energy use permits measurement of progress. It also permits rapid identification of unexpected increases in energy use, potentially an indication that something is amiss. Ultimately, energy efficiency improvements should show up in the monthly utility bills.

Report Success. Continuous tracking of energy use will also permit periodic reporting of success. This is critical in keeping building occupants engaged in the process.

Reassess Process and Goals. Energy efficiency improvement is a multiyear process. Conditions change over time and this may require reassessments of and adjustments to the energy efficiency plan. Factors that contribute to the need for reassessment include:

- Large changes in energy costs
- Large changes in the cost differential between electricity and fossil fuels
- Changes in the financial condition of the building as a profit center
- Changes in building occupancy and operating hours
- New energy efficiency technologies

DEFINING AND REPORTING SUCCESS

Over time, the cost of energy will continue to rise. Energy savings may result in decreases in energy costs in the short term, but in the long term, most of the savings will result in avoided energy cost. While economics drive energy efficiency decisions, absolute progress must be measured by tracking energy use rather than cost. Energy savings can be reported as a percent improvement against a base year. Avoided cost for the year can be calculated as savings for each energy type at current costs.

Three types of adjustments can be considered when comparing energy use in various years:

- Weather adjustments—year to year variations in weather can mask energy efficiency changes in the building
- Occupancy—changes in building occupancy increase or decrease energy use
- Plug loads—increases in equipment of all kinds will increase energy use along with internal heat gain

Reporting success is extremely important. Success requires effort from operating staff and cooperation from building occupants. Let them know their efforts are meaningful.

Appendix A: EUIs by Weather Zone

A

Variation in expected EUIs by ASHRAE weather zone has been estimated for presentation in this section. EUI goals can be set relative to the appropriate weather zone, as indicated on Figure A-1. Weather zone variations are based on a special analysis of the existing buildings set of the "DOE Commercial Reference Buildings" by the National Renewable Energy Laboratory (EERE 2011). Go to www1.eere.energy.gov for more information on the DOE commercial reference buildings.

In this special analysis, 16 building types were analyzed in 16 climate zones. Based on this analysis, EUIs were estimated for 48 of the building types in all 16 climate zones contained in the CBECS 2003 survey (EIA 2003) and also for the "average" for the entire country. This analysis is an extension of the work reported in *Methodology for Modeling Building Energy Performance across the Commercial Sector* (Griffith et al. 2008).

The 16 climate zones and representative cities used to calculate typical weather effects for the reference buildings are shown in Table A-1.

The 3B-coast zone is identified as 3B-c in the table of EUIs, and refers to the small region along the coast in region 3B. The estimated zonal EUIs are calculated as a ratio × "typical" EUI presented earlier in this book. The typical EUIs are called "U.S. EUI" in the table here. The "ratio" is the EUI for the representative city divided by the "average" for the country (Griffith et al. 2008).

Table A-1. Climate Zones—Representative Cities

Climate Zone	Representative City
1A	Miami, Florida
2A	Houston, Texas
2B	Phoenix, Arizona
3A	Atlanta, Georgia
3B-coast	Los Angeles, California
3B	Las Vegas, Nevada
3C	San Francisco, California
4A	Baltimore, Maryland
4B	Albuquerque, New Mexico
4C	Seattle, Washington
5A	Chicago, Illinois
5B	Boulder, Colorado
6A	Minneapolis, Minnesota
6B	Helena, Montana
7	Duluth, Minnesota
8	Fairbanks, Alaska

The results for these representative cities have been mapped to climate zones covering the entire country, as shown in Figure A-1. Using this map, the most appropriate value from the EUI table can be selected based on location. Note that the map has "moisture" divisions of Moist (A), Dry (B), and Marine (C). Table A-2 columns for the EUI values have A, B, and C identifiers for applicable zone numbers. It should also be noted that there is a wide variation in the energy use of some building types by climate zone.

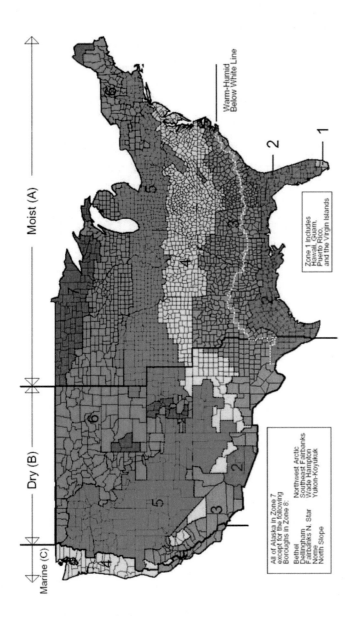

Figure A-1. Climate zone map (Briggs et al. 2002a, 2002b, 2002c).

Table A-2. EUIs Adjusted for Weather Zone

Estimated Zonal Site EUIs by Building Type, kBtu/ft^2 (ratio × U.S. EUI)

Building Use Description	U.S. EUI	1A	2A	2B	3A	3B-coast	3B	3C	4A	4B	4C	5A	5B	6A	6B	7	8
								ASHRAE Climate Zone									
Education																	
College/University (Campus-Level)					CBECS does not have data on college campuses.												
Elementary/Middle School	55	54	53	52	52	43	49	48	57	51	51	61	54	70	62	74	106
High School	78	68	70	67	69	50	65	60	83	70	74	92	78	108	93	118	171
Other Classroom Education	45	39	40	39	40	29	37	35	48	40	43	53	45	62	54	68	99
Preschool/Day Care	78	76	76	73	73	61	70	68	82	72	72	87	77	99	88	105	150
Food Sales																	
Grocery Store/Food Market	178	155	165	157	168	150	156	164	181	166	178	192	177	205	194	219	263
Convenience store (without Gas Station)	263	229	243	231	247	222	230	242	267	245	262	283	261	302	287	324	387
Convenience store (with Gas Station)	230	200	213	202	216	194	201	211	233	214	230	247	228	264	251	283	339
Other Food Sales	73	63	67	64	68	61	63	67	74	68	72	78	72	83	79	89	107
Food Service																	
Fast Food	419	380	389	382	398	352	384	372	432	402	408	466	428	506	470	543	673
Restaurant/Cafeteria	257	222	232	224	241	205	229	227	268	245	256	289	264	313	292	338	419

Table A-2. EUIs Adjusted for Weather Zone *(continued)*

Estimated Zonal Site EUIs by Building Type, kBtu/ft^2 (ratio × U.S. EUI)

Building Use Description	U.S. EUI	1A	2A	2B	3A	3B-coast	3B	3C	4A	4B	4C	5A	5B	6A	6B	7	8
						ASHRAE Climate Zone											
Other Food Service	147	127	132	128	138	117	131	130	153	140	146	165	151	179	166	193	239
Health Care																	
Hospital/Inpatient Health	198	199	202	190	195	188	186	194	203	175	190	203	179	210	188	213	254
Nursing Home, Assisted Living	126	120	119	115	118	99	111	107	130	117	120	141	126	156	142	168	223
Clinic/Other Outpatient Health	72	74	73	73	72	67	73	63	73	72	65	71	71	74	72	73	85
Medical Office (non-diagnostic)	43	40	41	40	39	32	36	36	43	37	40	47	40	52	46	54	73
Medical Office (diagnostic)	44	46	45	45	45	41	45	39	45	45	40	44	44	46	45	45	53
Laboratory	266	255	252	244	250	210	236	227	277	247	255	299	267	331	302	356	473
Lodging																	
Hotel	76	61	67	62	72	65	65	70	79	74	77	86	81	93	90	101	122
Motel or Inn	73	70	70	68	69	61	66	63	74	69	68	79	73	86	79	91	111
Dormitory/Fraternity/Sorority	74	70	70	69	69	57	66	59	77	68	69	86	75	99	88	108	138
Other Lodging	71	67	67	66	67	59	64	61	71	66	66	76	70	83	76	88	107

Table A-2. EUIs Adjusted for Weather Zone *(continued)*

Estimated Zonal Site EUIs by Building Type, kBtu/ft^2 (ratio × U.S. EUI)

ASHRAE Climate Zone

Building Use Description	U.S. EUI	1A	2A	2B	3A	3B-coast	3B	3C	4A	4B	4C	5A	5B	6A	6B	7	8
Mall																	
Strip Mall	94	78	82	80	88	62	80	74	105	89	97	119	101	139	125	156	219
Enclosed Mall	94	78	82	80	87	62	80	74	104	89	97	118	101	138	124	155	218
Office																	
Administrative/ Professional Office	67	62	64	62	60	50	56	55	67	58	61	72	62	79	70	83	113
Bank/Financial Institution	89	82	85	82	81	67	75	73	89	77	81	96	83	106	94	111	151
Government Office	77	71	73	70	69	57	64	63	76	66	70	82	71	91	80	95	129
Mixed-Use Office	78	72	74	72	70	58	65	64	78	67	71	83	72	92	82	97	131
Other Office	59	55	56	55	53	44	49	48	59	51	54	64	55	70	62	74	100
Other (all other types, which is a very wide range)	70	climate zone variation not estimated for "Other" category															
Public Assembly																	
Entertainment/ Culture	46	44	44	42	43	36	41	39	48	43	44	52	46	57	52	62	82
Library	94	90	89	86	88	74	83	80	98	87	90	105	94	117	107	126	167
Recreation	48	46	46	44	45	38	43	41	50	45	46	54	48	60	55	65	86

Table A-2. EUIs Adjusted for Weather Zone *(continued)*

Estimated Zonal Site EUIs by Building Type, kBtu/ft^2 (ratio × U.S. EUI)

ASHRAE Climate Zone

Building Use Description	U.S. EUI	1A	2A	2B	3A	3B-coast	3B	3C	4A	4B	4C	5A	5B	6A	6B	7	8
Social/Meeting	59	56	55	54	55	46	52	50	61	54	56	66	59	73	66	78	104
Other Public Assembly	42	40	40	39	40	33	37	36	44	39	40	47	42	52	48	56	75
Public Order and Safety																	
Fire Station/Police Station	98	94	93	90	92	77	87	84	102	91	94	110	98	122	111	131	174
Courthouse	93	89	88	85	87	73	82	79	97	87	89	104	93	116	106	125	165
Religious Worship	38	36	35	34	35	30	33	32	39	35	36	42	38	47	43	50	67
Retail																	
Retail Stores (non-mall stores)	48	43	43	41	42	31	38	34	50	42	45	56	48	64	58	72	100
Other Retail	92	82	83	79	81	59	74	66	95	80	86	108	91	123	110	137	193
Vehicle Dealerships/Showrooms	82	74	74	71	73	53	66	59	85	72	77	97	82	111	99	123	173
Service																	
Vehicle Repair/Service Shop	43	41	41	39	40	34	38	37	45	40	41	48	43	53	49	57	76
Vehicle storage/maintenance	28	26	26	25	26	22	24	23	29	26	26	31	28	34	31	37	49
Post Office/Postal Center	71	67	67	65	66	56	62	60	73	66	68	79	71	88	80	94	125

Table A-2. EUIs Adjusted for Weather Zone (continued)

Estimated Zonal Site EUIs by Building Type, kBtu/ft^2 (ratio × U.S. EUI)

Building Use Description	U.S. EUI	1A	2A	2B	3A	3B-coast	3B	3C	4A	4B	4C	5A	5B	6A	6B	7	8
							ASHRAE Climate Zone										
Repair Shop	48	46	46	44	45	38	43	41	50	45	46	54	48	60	55	65	86
Other Service	90	86	85	82	85	71	80	77	94	84	86	101	90	112	102	120	160
Storage/Shipping/ Warehouse																	
Self-storage	7				climate zone variation not estimated for "Self-Storage" category												
Nonrefrigerated Warehouse	19	28	18	18	17	14	17	14	20	18	17	23	21	27	25	31	49
Distribution/Shipping Center	34	49	31	32	31	24	30	25	35	32	30	40	38	48	44	55	86
Refrigerated Warehouse	126	121	119	116	119	99	112	108	131	117	121	142	127	157	143	169	224
Vacant	11				climate zone variation not estimated for "Vacant" category												

Source:
U.S. EUI (median EUI) calculated based on DOE/EIA 2003 CBECS microdata with malls (EIA 2003). Climate variations based on special analysis of simulations of the "DOE reference buildings" (EERE 2011) and resulting ratios of the climate EUI to the weighted average EUI.

Notes:
1. Propane calculations based on methodology similar to ENERGY STAR (EPA 2008), with some deletions based on uncertainties in propane use.
2. Buildings larger than one million square feet have inappropriate data and were deleted.
3. Except for vacant buildings, any building used less than nine months/yr was deleted.
4. Malls do not have the months of use variable coded, so malls were not included in the months-of-use criterion.

Appendix B: Setting Energy Goals—Examples   B

Setting energy target goals and determining how a building uses energy is often limited to the availability of data. Annual and monthly energy use from utility bills is the most prevalent data and provides EUI and ECI metrics for benchmark comparisons and target setting. Total energy-use goals are often set to reach some threshold compared to a benchmark or a percentage change from historical use. Energy data at finer time resolution provides insight into when energy is used and allows goals to be set for energy use at specific times of the day (e.g., unoccupied energy use). Submetered data provide insight into system performance and allow goals to be set on a system level basis. The additional information available from more detailed data support setting goals related to specific systems or time periods. For example, one might try and limit unoccupied energy use by reviewing an inventory of equipment that might operate at night. The availability of interval data helps define the issue and provides feedback on implementation. The following data examples illustrate how various levels of data can be used to review performance and set goals.

UTILITY BILLING DATA FOR SETTING HISTORICAL PERFORMANCE GOALS

Utility bills are readily available and provide a time series of monthly energy use that can also be correlated to weather. In the following example, the building has no cooling. Electricity use is fairly constant throughout the year with a slight decline in later years reaching 50,000 kWh/month at times where previously 60,000 kWh/month was the lowest observed energy use. In setting a goal from monthly time series data, one needs to be able to smooth or account for typical variations in the monthly energy use (e.g., meter-reading frequency, weather, etc). In the data in Figure B-1, monthly energy use was ranging from 60,000–75,000 kWh/month in 2006 and 2007. This range decreased to 50,000–65,000 kWh/month in 2009 and 2010.

The same data is shown relative to ambient weather in Figure B-2. The building uses fossil fuels heat, so the slight increase in electricity use with colder weather could indicate additional fan power, additional lighting from fewer daylight hours, or occupant space heater use. Similarly, the lack of mechanical air-conditioning might lead to more occupant fans, resulting in a slight increase in electricity use with warmer weather.

Reviewing the monthly natural gas use shows a typical pattern for a northern climate building with gas heat. Of note is that the summer gas use shows the non-weather-related loads for service water heating and cooking. The weather- and non-weather-related gas uses can be identified by inspecting the monthly gas use in Figure B-3.

The gas data become more useful when related to ambient temperature in Figure B-4. The consistent slope of the heating load line gives a signature for the building gas use. The slope is related to the thermal performance of the building (heat conduction of walls, windows, and roof), volume of outdoor air introduced to the building (ventilation rates, infiltration), and the efficiency of the heating plant. The amount of heat gain from occupants, equipment, and lighting also influence the load line. The outdoor temperature where the heating load balances the internal gains is the thermal balance point, denoted by the temperature at which the gas use begins to increase with decreasing ambient temperatures. In the following example, the balance temperature is 65°F, indicating either poor thermal performance of the building envelope or very low internal gains. Typically, commercial buildings have thermal balance points in the 50s. A target or goal for the heating load slope can be derived from a building energy model or calculations of heat loss and outdoor air flow rates.

All of these analyses were performed in a spreadsheet with utility bill and weather data. Some utilities provide heating degree day and/or average ambient information for the billing period that can also be used to represent weather conditions.

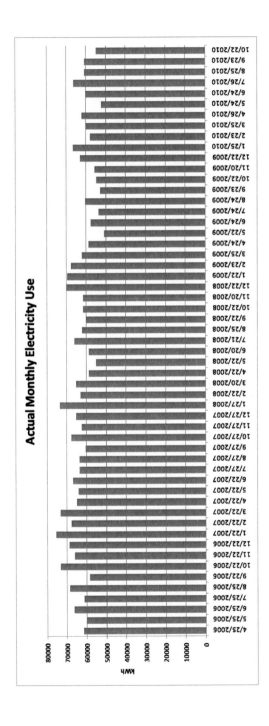

Figure B-1. Actual monthly energy use.

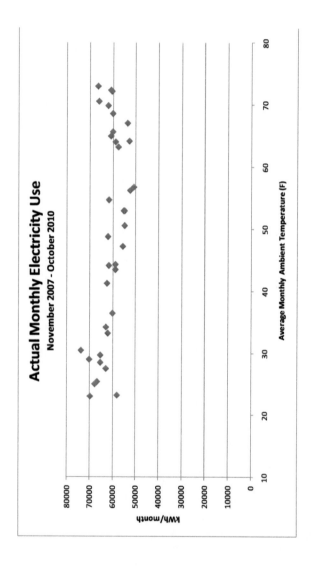

Figure B-2. Actual monthly electricity use by ambient temperature.

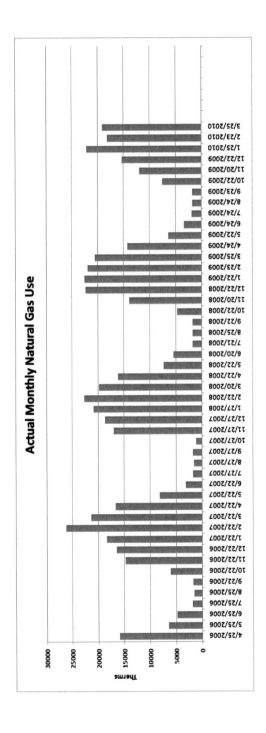

Figure B-3. Actual monthly gas use.

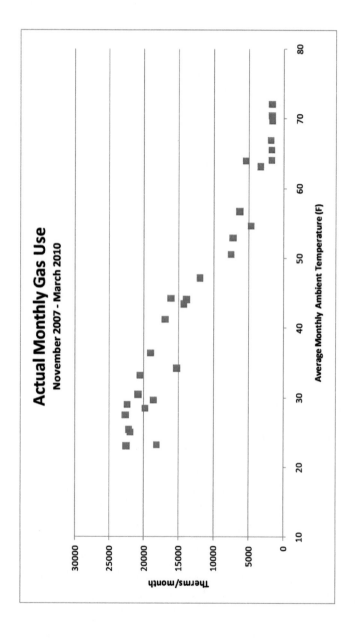

Figure B-4. Actual monthly gas use over a three-year period.

INTERVAL ENERGY DATA FOR SETTING TEMPORAL PERFORMANCE GOALS

Similar analysis can be performed on data available at higher frequency intervals. Daily total energy use from submetered or manually read billing meters provide additional refinement to the heating load line. The weekday and weekend differences in energy use can be quantified. Also, interval meter data provide a pulse from the building showing variation in energy use throughout the day, allowing assessment of unoccupied energy use levels and confirmation of building operation schedules from energy using equipment. These types of data can support goals of driving down unoccupied energy use to a certain level, or limiting peak demand at startup through staggering system startups in the morning. In the following examples, Figures B-5 and B-6 show the 15-minute total building power from an electrically heated and cooled building in a heating month and a cooling month. Nighttime power levels of 20–30 kW give targets for investigating equipment operation during unoccupied periods so that one knows when he or she is close to finding most operating equipment. Target levels can then be set and verified.

The time-series data facilitates the comparison of operation and upgrades toward meeting a target. In the following example building, a lighting and motor retrofit was undertaken. Pre- and post-retrofit data show the impact of the investments. Reduction in the constant lighting and motor power created a consistent offset in the pre- and post-retrofit monitoring period. The long-term time-series plots shown in Figures B-7 and B-8, show the progression in energy reduction as the project was implemented between February and April.

Daily data allow identification of energy use on different day types. Figure B-9 shows the pre- and post-retrofit periods and distinguishes weekends from weekdays.

When submetered data are available, it's possible to set and review energy performance goals on a system level basis. Figure B-10 shows cooling, pump, and fan energy expectations.

BENCHMARK DATA FOR PERFORMANCE COMPARISONS

Various sources for comparison might be possible depending on the building type. Raw EUI comparisons can be made from national data. Comparisons of building EUIs are possible, normalized for both weather and building use. An organization with a portfolio of buildings has a comparative source of building data for benchmarking.

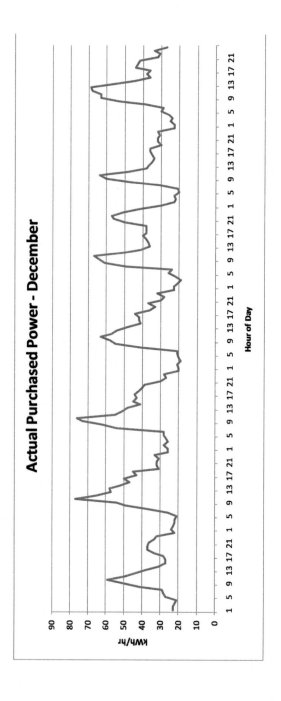

Figure B-5. Actual purchased power in December.

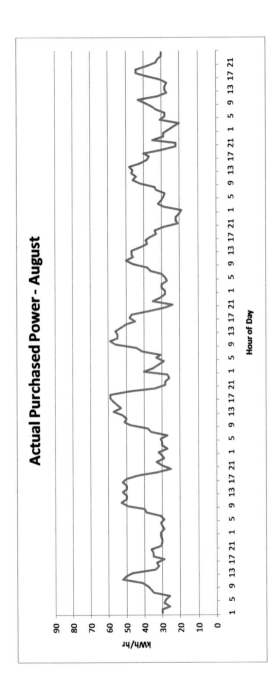

Figure B-6. Actual purchased power in August.

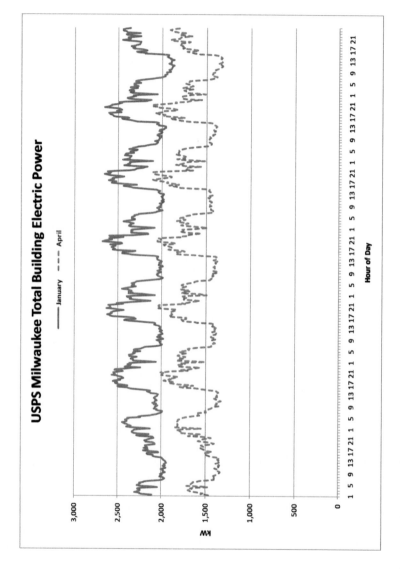

Figure B-7. Total building electric power.

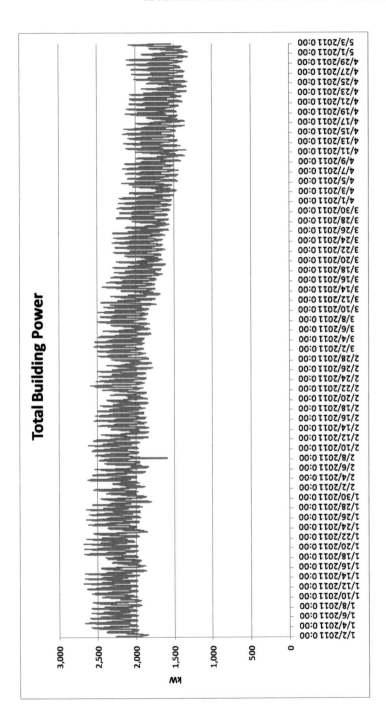

Figure B-8. Time series electric power.

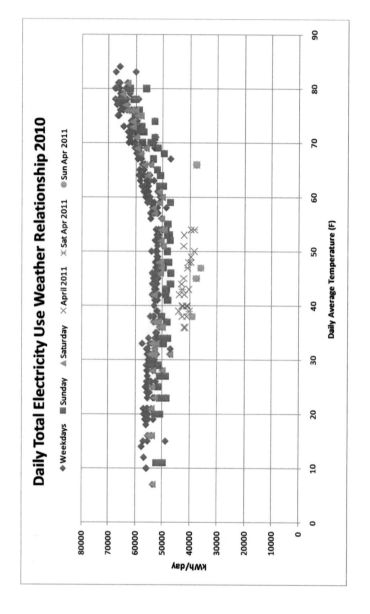

Figure B-9. Energy use for weekdays versus weekends.

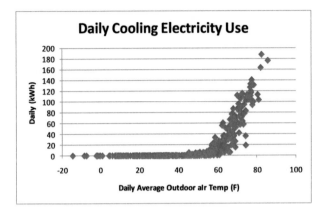

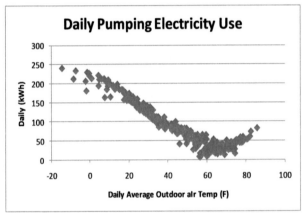

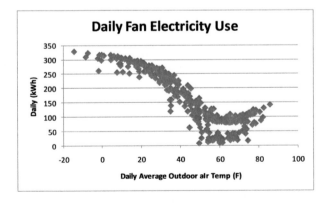

Figure B-10. Daily cooling, fan, and pump electricity use.

Table B-1. Weighted Electricity Use by Building Type from 2003 CBECS (EIA 2003)

| Building Use | Weighted Electricity Use Index Values, kWh/yr per gross ft^2 | | | | | |
| | Percentiles | | | | | |
	10th	25th	50th	75th	90th	Mean
Administrative/ Professional office	3.54	6.7	11.0	15.0	24.1	12.7
Bank/Other financial	6.23	14.5	22.2	29.5	33.3	22.5
Clinic/Other outpatient health	4.94	9.4	15.2	20.7	27.3	16.6
College/University	4.13	10.5	15.0	24.0	42.3	17.7

The example in Table B-1 shows the electricity use on a floor area basis with values across the distribution of observations by building type. If your building falls into one of the reported categories, the national energy data can give some rough guidance as to where the energy use is relative to similar building. This type of raw data comparison ignores climate and building use characteristics and is therefore only suited for a high-level rough estimate of relative energy use evaluation.The ENERGY STAR Portfolio Manager provides a benchmarking system normalizing energy use for building use and climate. The system is described in the main body of the book as a popular benchmarking tool (EPA 2008a).

An internal reference is possible when an organization has multiple facilities that share a similar function. Retail stores, supermarkets, and schools are all examples. If these facilities are in a geographically similar region there might not be a need to normalize for weather as all operate on the same conditions. A simple comparison of how performance metrics rank can provide insight into which facilities are consistently performing better. The goal is often to try and shrink the difference between the best and worst performing by improving the facilities with the lowest performance metrics. As with historical performance evaluations, the internal reference doesn't have a means to compare with other facilities.

The best-performing building in a portfolio might not necessarily compare well with a similar building of other organizations or compare well to minimum energy standards.

In the example shown in Figures B-11 and B-12, the distribution in the organization's ECIs for electricity and gas show a large range for similar facilities in a close geographical region (similar weather). Electricity costs range mostly from \$2–\$4/ft^2 while natural gas costs range from about \$0.20–\$1.00/ft^2. An investigation of the characteristics of the outliers can be useful in identifying potential performance issues. This type of comparison can also provide an idea of the financial impact of improving energy performance within the portfolio. Reducing the top highest electricity and natural gas using facilities to the level of the 75th percentile (\$3.30/ft^2 electricity and \$0.60/ft^2 gas) results in \$600,000/year in electricity and \$300,000/year in gas savings. This type of portfolio review not only helps identify which facilities need attention, but illustrates to management the potential financial impact if the facilities performed more in line with the norm.

PERFORMANCE COMPARISONS TO THEORETICAL TARGETS

An energy model can provide a theoretical target for energy performance for the entire building as well as for individual systems that are explicitly modeled. Annual, monthly, daily, and shorter time period energy-use data can be produced from the model. The model also offers the flexibility to quantify the impacts of changes or deficiencies. The model can be run with and without system energy efficiency features, such as energy-recovery ventilation, or variable-speed pumps and fans to quantify the impact that might be observed in actual energy-use data. Figure B-13 shows actual vs. modeled total building electricity-use variation with outdoor air temperature.

In Figure B-14, deviations from these trends can be representative of improvements or problems. The second graphic in each row depicts the ideal modeled energy use for the systems as a comparison to expectations. The most notable difference is the larger measured daily gas use trend compared to the modeled trend. This deviation was tracked down to energy recovery ventilation equipment that was not operating, causing an increased heating load to condition the outdoor air. The measured data figures also illustrate the daily variation in energy use. A change would need to be relatively large to appear outside the "normal" noise in the daily energy use.

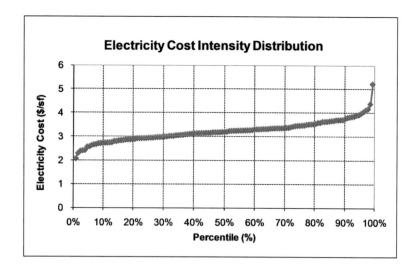

Figure B-11. Electricity cost intensity rank.

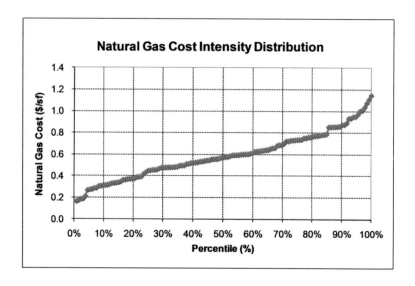

Figure B-12. Natural gas cost intensity rank.

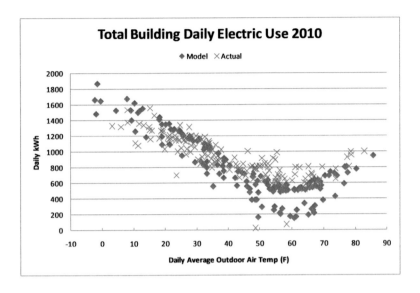

Figure B-13. Total daily electricity use—modeled versus actual.

SUMMARY

Goal setting can be relative to past history or compared to benchmarks. The availability of both measured and comparative data shape the types of comparisons and types of goals that can be considered. Common data types include:

- Monthly utility bills
- Daily total building energy use (manual reads, data logging)
- Total building interval data (hourly, 15-minute)
- End-use and/or submetered energy data

These data can be compared to various sources:

- Historical building data
- Similar building in an organization's portfolio
- External data sources of similar building data
- Benchmark systems (often with normalizing factors)
- Theoretical performance targets (engineering calculations and energy models)

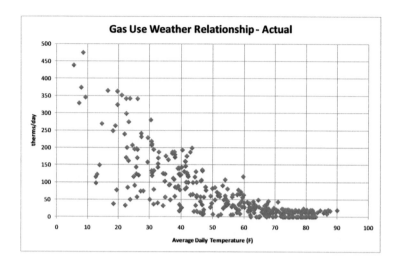

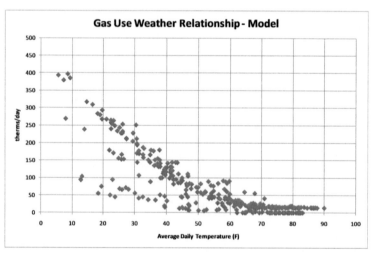

Figure B-14. a) Daily natural gas use by day type and outdoor air temperature.

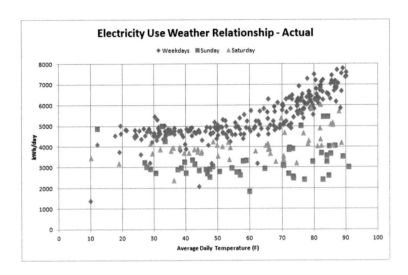

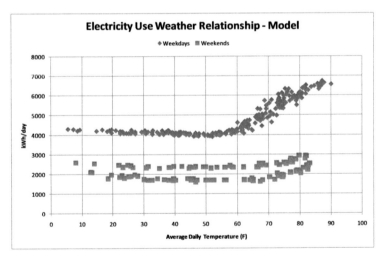

Figure B-14. b) Daily electricity use by day type and outdoor air temperature.

References
and Web Sites

Abate, D., M. Towers, R. Dotz, L. Romani, and P. Lufkin. 2010. *The Whitestone Facility Maintenance and Repair Cost Reference*, 15th ed. Washington, DC: Whitestone Research. www.whitestoneresearch.com/research/news/annual-cost-reference-update-2010.aspx.

Abramson, B., D. Herman, and L. Wong. 2005. Interactive Web-based owning and operating cost database. Research Project 1237, American Society of Heating, Refrigerating and Air-Conditioning Engineers, Atlanta.

ASHRAE. 1995. *ASHRAE Handbook—HVAC Applications*, Chapter 49, Table 5. Atlanta: American Society of Heating, Refrigerating, and Air-Conditioning Engineers.

ASHRAE. 2002. *ASHRAE Guideline 14-2002, Measurement of Energy and Demand Savings*, Atlanta: American Society of Heating, Refrigerating, and Air-Conditioning Engineers.

ASHRAE. 2005. *ASHRAE Guideline 0-2005, The Commissioning Process*, Atlanta: American Society of Heating, Refrigerating, and Air-Conditioning Engineers.

ASHRAE. 2006a. *ASHRAE Guideline 1.2P: The Commissioning Process for Existing HVAC&R Systems*, Atlanta: American Society of Heating, Refrigerating, and Air-Conditioning Engineers.

ASHRAE. 2006b. *ASHRAE Guideline 1.3P: Building Operation and Maintenance Training for the HVAC&R Commissioning Process*, Atlanta: American Society of Heating, Refrigerating, and Air-Conditioning Engineers.

ASHRAE. 2007a. ANSI/ASHRAE Standard 52.2-2007, *Methods of Testing General Ventilation Air-Cleaning Devices for Removal Efficiency by Particle Size*. Atlanta: American Society of Heating, Refrigerating, and Air-Conditioning Engineers.

ASHRAE. 2007b. ANSI/ASHRAE Standard 105-2007, *Standard Methods of Measuring, Expressing and Comparing Building Energy Performance*. Atlanta: American Society of Heating, Refrigerating, and Air-Conditioning Engineers.

ASHRAE. 2007c. *ASHRAE Guideline 1.1: HVAC&R Technical Requirements for the Commissioning Process*, Atlanta: American Society of Heating, Refrigerating, and Air-Conditioning Engineers.

ASHRAE. 2008a. *ASHRAE Guideline 0.2P: The Commisioning Process for Existing Systems and Assemblies*, Atlanta: American Society of Heating, Refrigerating, and Air-Conditioning Engineers.

ASHRAE. 2008b. *ASHRAE Handbook—HVAC Systems and Equipment*, Chapter 24. Atlanta: American Society of Heating, Refrigerating, and Air-Conditioning Engineers.

ASHRAE. 2009a. ANSI/ASHRAE/USGBC/IES Standard 189.1-2009, *Standard for the Design of High-Performance Green Buildings.* Atlanta: American Society of Heating, Refrigerating, and Air-Conditioning Engineers.

ASHRAE. 2009b. *ASHRAE Guideline 1.4P: Systems Manual Preparation for the Commissioning Process*, Atlanta: American Society of Heating, Refrigerating, and Air-Conditioning Engineers.

ASHRAE. 2009c. *ASHRAE Handbook—Fundamentals*, Chapter 36. Atlanta: American Society of Heating, Refrigerating, and Air-Conditioning Engineers.

ASHRAE. 2009d. *The Indoor Air Quality Guide: Best Practices for Design, Construction and Commissioning*. Atlanta: American Society of Heating, Refrigerating, and Air-Conditioning Engineers.

ASHRAE. 2010a. ANSI/ASHRAE Standard 55-2010, *Thermal Environmental Conditions for Human Comfort.* Atlanta: American Society of Heating, Refrigerating, and Air-Conditioning Engineers.

ASHRAE. 2010b. ANSI/ASHRAE Standard 62.1-2010, *Ventilation for Acceptable Indoor Air Quality.* Atlanta: American Society of Heating, Refrigerating, and Air-Conditioning Engineers.

ASHRAE. 2010c. ANSI/ASHRAE/IES Standard 90.1-2010, *Energy Standard for Buildings Except Low-Rise Residential Buildings.* Atlanta: American Society of Heating, Refrigerating, and Air-Conditioning Engineers.

ASHRAE. 2010d. ASHRAE Owning and Operating Cost Database, *Equipment life/maintenance cost survey.* www.ashrae.org/database.

ASHRAE. 2010e. *Performance Measurement Protocols for Commercial Buildings.* Atlanta: American Society of Heating, Refrigerating, and Air-Conditioning Engineers.

ASHRAE. 2011a. Advanced Energy Design Guide Series. Atlanta: American Society of Heating, Refrigerating, and Air-Conditioning Engineers. www.ashrae.org/publications/page/1604.

ASHRAE. 2011b. *ASHRAE Handbook—HVAC Applications,* Chapter 36. Atlanta: American Society of Heating, Refrigerating, and Air-Conditioning Engineers.

ASHRAE. 2011c. *ASHRAE Handbook—HVAC Applications,* Chapter 37. Atlanta: American Society of Heating, Refrigerating, and Air-Conditioning Engineers.

ASHRAE. 2011d. *Procedures for Commercial Building Energy Audits*, 2d ed. Atlanta: American Society of Heating, Refrigerating, and Air-Conditioning Engineers. Forthcoming 2011.

ASTM. 2010. ASTM Standard C518-10, *Standard Test Method for Steady-State Thermal Transmission Properties by Means of the Heat Flow Meter Apparatus.* American Society for Testing and Materials, West Conshohocken, PA.

BH. 2010. Insulated Roof Yields Energy Savings at Springboard Engineering. Black Hills Energy, Rapid City, SD. www.blackhillsenergy.com/services/programs/documents/ia_case_study_sprbrd.pdf.

BOMA. 2010. Experience Exchange Report (EER). Building Owners and Managers Association International, Washington, DC. www.boma.org/resources/benchmarking/Pages/default.aspx.

BOMA. 2011. *BOMA International Commercial Lease: Guide to Sustainable and Energy Efficient Leasing for High-Performance Buildings* (formerly *Green Lease Guide*). Washington, DC: Building Owners and Managers Association International.

Boyce, P.R., Eklund, N.H., and Simpson, S.N. 1999. Individual lighting control: Task performance, mood, and illuminance. Report, Lighting Research Center, Rensselaer Polytechnic Institute, Watervliet, NY. www.lrc.rpi.edu/resources/pdf/67-1999.pdf.

Briggs, R.S., R.G. Lucas, and Z.T. Taylor. 2002a. Climate classification for building energy codes and standards, PNNL Technical Paper final review draft. Richland, WA: Pacific Northwest National Laboratory.

Briggs, R.S., R.G. Lucas, and Z.T. Taylor. 2002b. Climate classification for building energy codes and standards: Part 1—Development process. *ASHRAE Transactions* 109 (1).

Briggs, R.S., R.G. Lucas, and Z.T. Taylor. 2002c. Climate classification for building energy codes and standards: Part 2—Zone definitions, maps, and comparisons. *ASHRAE Transactions* 109 (1).

CEC. 2000. *Energy Efficiency Project Handbook: How to Hire an Energy Auditor to Identify Energy Efficiency Projects.* Sacramento, CA: California Energy Commission. www.energy.ca.gov/reports/efficiency_handbooks/400-00-001C.pdf.

CEC. 2011. *Summertime Energy-Saving Tips For Businesses.* California Energy Commission Consumer Energy Center. www.consumerenergycenter.org/tips/business_summer.html.

CEP. 2011. The Climate and Energy Project, *Tips for Farms and Businesses.* www.climateandenergy.org/TakeStep/TipsForFarmsAndBusinesses/Index.html.

Choi, J.M. and Kim, Y.C. 2001. The Effects of Improper Refrigerant Charge on the Performance of a Heat Pump with an Electronic Valve and Capillary Tube. Technical Report, Department of Mechanical Engineering, Korea University, Seoul, Korea.

CIBO. 1997. *CIBO Energy Efficiency Handbook.* Burke, VA: Council of Industrial Boiler Owners. www.cibo.org/pubs/steamhandbook.pdf.

Con Edison. 2009. Consolidated Edison Company of New York, Inc. *Rates and Tariffs: Schedule for Steam Service.* www.coned.com/rates/steam-sched.asp.

DOE, HI, and Europump. 2004. Variable Speed Pumping—A Guide to Successful Applications, Executive Summary. Washington, DC: U.S. Department of Energy, Office of Industrial Technology. Parsippany, NJ: Hydraulic Institute. Brussels, Belgium: Europump. www1.eere.energy.gov/industry/bestpractices/pdfs/variable_speed_pumping.pdf.

DOE. 2010. *MotorMaster+,* Version 4.01.01. U.S. Department of Energy, Office of Industrial Technology, Washington, DC. www1.eere.energy.gov/industry/bestpractices/software_motormaster.html.

DOE. 2011a. Determining Electric Motor Load and Efficiency, DOE/GO-10097-517, U.S. Department of Energy, Washington, DC. www.scribd.com/doc/54779818/Electric-Motor-Load-Efficiency.

DOE. 2011b. Lighting Facts®, *Success with solid-state lighting.* U.S. Department of Energy. www.lightingfacts.com.

EERE. 2008a. Insulation Fact Sheet. Prepared by Oak Ridge National Laboratory for U.S. Department of Energy, Office of Energy Efficiency and Renewable Energy, Washington, DC. www.ornl.gov/sci/roofs+walls/facts/Insulation%20Fact%20Sheet%202008.pdf.

EERE. 2008b. *M&V Guidelines: Measurement and Verification for Federal Energy Projects,* version 3. Washington, DC: U.S. Department of Energy, Office of Energy Efficiency and Renewable Energy. www1.eere.energy.gov/femp/pdfs/mv_guidelines.pdf.

EERE. 2011a. Commerical Reference Buildings, Commercial Building Initiative. Washington, DC: U.S. Department of Energy, Energy Efficiency and Renewable Energy. www1.eere.energy.gov/buildings/commercial_initiative/reference_buildings.html.

EERE. 2011b. U.S. Department of Energy, Office of Energy Efficiency and Renewable Energy. www1.eere.energy.gov.

EIA. 2003. *2003 Commercial Buildings Energy Consumption Survey.* Washington, DC: Energy Information Administration, U.S. Department of Energy. www.eia.gov/emeu/cbecs.

ENVEST. 1985. *Energy Efficiency Investment Analysis Software Manual.* Commonwealth of Pennsylvania: Alliance to Save Energy and Governor's Energy Council.

EPA. 1991. *Building Air Quality: A Guide for Building Owners and Facility Managers.* Washington, DC: U.S. Environmental Protection Agency. www.epa.gov/iaq/largebldgs/pdf_files/iaq.pdf.

EPA. 2007. Putting Energy Into Profits: ENERGY STAR® Small Business Online Guide. U.S. Department of Energy, Washington, DC. www.energystar.gov/ia/business/small_business/sb_guidebook/smallbizguide.pdf.

EPA. 2008a. *ENERGY STAR Portfolio Manager.* Washington, DC: U.S. Environmental Protection Agency. www.energystar.gov/index.cfm?c=evaluate_performance.bus_portfoliomanager.

EPA. 2008b. *ENERGY STAR Performance Ratings: Methodology for Incorporating Source Energy Use.* Washington, DC: U.S. Environmental Protection Agency. www.energystar.gov/ia/business/evaluate_performance/site_source.pdf.

EPA. 2011a. ENERGY STAR: Cool Roofs and Emissivity. Washington, DC: U.S. Environmental Protection Agency.www.energystar.gov/index.cfm?c=roof_prods.pr_roof_emissivity.

EPA. 2011b. ENERGY STAR for Small Businesses. U.S. Department of Energy, Washington, DC. www.energystar.gov/index.cfm?c=small_business.sb_index.

EPA. 2011c. ENERGY STAR Target Finder. U.S. Department of Energy, Washington, DC. www.energystar.gov/index.cfm?c=new_bldg_design.bus_target_finder.

EPMI. 2011. Site fuel I-P units conversion calculator, Energy Performance Measurement Institute. http://epminst.us/industrial/IP-unit%20site%20fuel%20to%20energy%20calculator%20for%20multiple%20fuels.xls.

EVO. 2007. *International Performance Measurement and Verification Protocol.*, Vol. 3. Washington, DC: Efficiency Valuation Organization. www.evo-eorld.org.

Farzad, M. and O'Neal, D.L. 1988. An evaluation of improper refrigerant charge on the performance of a split system air-conditioner with capillary tube expansion. Technical Report 259, Energy Systems Laboratory, Texas A&M University, College Station, TX.

GA. 2006. Energy Use Simulation and Economic Analysis, *Table 4.6: Equipment Service Life*. Washington State Department of General Administration. www.ga.wa.gov/eas/elcca/simulation.html.

Goldner, F.S. 1999. DHW Recirculation system control strategies. Final Report 99-1, Prepared by Energy Management & Research Associates for New York State Energy Research and Development Authority, New York.

Green, C.H. 2007. Trust in business: The core concepts. *Trusted Advisor*. http://trustedadvisor.com/articles/Trust-in-Business-The-Core-Concepts.

Griffith, B., N. Long, P. Torcellini, R. Judkoff, D. Crawley, and J. Ryan. 2008. Methodology for modeling building energy performance across the commercial sector. Technical Report NREL/TP-550-41956, National Renewable Energy Laboratory/U.S. Department of Energy, Washington, DC. www.nrel.gov/docs/fy08osti/41956.pdf.

Griggs, E.I., T.R. Sharp, and J.M. Macdonald. 1989. Guide for estimating differences in building heating and cooling energy due to changes in solar reflectance of a low-sloped roof. Oak Ridge National Laboratory report. ORNL-6527, Oak Ridge National Laboratory, Oak Ridge, TN. http://epminst.us/otherEBER/ornl6527.pdf.

IES. 2000. *The IESNA Lighting Handbook: Reference and Application,* 9th ed. New York: Illuminating Engineering Society of North America.

IES. 2004. ANSI/IESNA RP-1-2004, *American National Standard Practice for Office Lighting.* New York: Illuminating Engineering Society of North America.Bethesda, MD: National Electrical Contractors Association.

IES. 2006. NECA/IESNA-502-2006, *American National Standard for Installing Industrial Lighting Systems.* New York: Illuminating Engineering Society of North America.

IES. 2011a. *The Commissioning Process Applied to Lighting and Control Systems.* DG-29-11, New York: Illuminating Engineering Society of North America.

IES. 2011b. *The IESNA Lighting Handbook: Reference and Application,* 10th ed. New York: Illuminating Engineering Society of North America.

IES. 2011c. *Luminaire Classification System for Outdoor Luminaires.* New York: Illuminating Engineering Society of North America.

Landsberg, D., M. Lord, S. Carlson, and F. Goldner. 2009. *Energy Efficiency Guide for Existing Commercial Buildings: The Business Case for Building Owners and Managers.* Atlanta: American Society of Heating, Refrigerating, and Air-Conditioning Engineers.

LBNL. 1993. *DOE-2,* Version 2.1e. Lawrence Berkeley National Laboratory in partnership with James J. Hirsch & Associates.

LBNL. 2011. Utility Energy Efficiency Programs for Small & Large Businesses. Lawrence Berkeley National Laboratory, Environmental Energy Technologies Division. http://eetd.lbl.gov/EnergyCrossroads/2ueeprogram.html.

McRae, M., M. Rufo, and D. Baylon. 1987. Service Life of Energy Conservation Measures. Final Report, prepared by XENERGY Inc. and Ecotope Inc. for the Bonneville Power Administration, Portland, OR.

NAESCO. 2011. Resources: Other sites of interest. www.naesco.org/resources/sites.htm. Washington, DC: National Association of Energy Services Companies.

National Grid. 2011. National Grid Energy Efficiency Programs in New York State. www.powerofaction.com/efficiency.

NCSU. 2011. Database of State Incentives for Renewable Energy (DSIRE). North Carolina State University, Raleigh, NC. www.dsireusa.org.

NEMA. 2010. NEMA Standard MG-2009, *Motors and Generators*. National Electrical Manufacturers Association, Rosslyn, VA. www.nema.org/prod/ind/motor.

NYSERDA. 2011a. Energy Audit Program: Lower Energy Bills and Improved Energy Performance. http://nyserda.ny.gov/Page-Sections/Commercial-and-Industrial/Programs/FlexTech-Program/Energy-Audit-Program.aspx.

NYSERDA. 2011b. Existing Facilities Program. www.nyserda.ny.gov/Page-Sections/Commercial-and-Industrial/Programs/Existing-Facilities-Program.

ORNL/LBNL. 2010. *Roof Savings Calculator (RSC)*, Beta Release v0.92. Oak Ridge National Laboratory, Oak Ridge, TN, Lawrence Berkeley National Laboratory, Berkeley, CA. www.roofcalc.com.

PECI. 2009. A study on energy savings and measure cost effectiveness of existing building commissioning. Portland Energy Conservation, Portland, OR. www.peci.org/documents/annex_report.pdf.

SBA. 2011. U.S. Small Business Administration, *Easy Energy Efficiency Improvements*. www.sba.gov/content/easy-energy-efficiency-improvements.

T.C. Chan Center for Building Simulation Studies. 2008. *Impact of Shades on Solar Heat Gain*. http://tcchancenter.com/research/fundamental-studies/material-research/green-material-selection-guide.

USGBC. 2008. *Leadership in Energy and Environmental Design (LEED) for Existing Buildings: Operations & Maintenance*. Washington, DC: U.S. Green Building Council. www.usgbc.org/ShowFile.aspx?DocumentID=3617.

Wisconsin. 2011. Wisconsin's Focus on Energy: Small commercial building program. www.focusonenergy.com.

Web Sites:

ASHRAE Owning and Operating Cost Database
www.ashrae.org/database

California Energy Commission
www.energy.ca.gov

CBECS—Commercial Buildings Energy Consumption Survey
www.eia.gov/emeu/cbecs

CEC—California Energy Commission Consumer Energy Center
www.consumerenergycenter.org

CEP—The Climate and Energy Project
www.climateandenergy.org

DOE—U.S. Department of Energy
www1.eere.energy.gov

DSIRE—Database of State Incentives for Renewable Energy
www.dsireusa.org

Energy Crossroads at Lawrence Berkeley National Laboratory
http://eetd.lbl.gov/EnergyCrossroads/2ueeprogram.html

ENERGY STAR® Portfolio Manager
www.energystar.gov/index.cfm?c=evaluate_performance.bus
_portfoliomanager

LEED—Leadership in Energy and Environmental Design
www.usgbc.org

Lighting Facts®
www.lightingfacts.com

Universal Translator
www.utonline.org

U.S. Small Business Administration
www.sba.gov.